Lecture Notes in Mathematics

Edited by A. Dold and B. Eckmann

1154

D.S. Naidu
A.K. Rao

Singular Perturbation Analysis of Discrete Control Systems

Springer-Verlag
Berlin Heidelberg New York Tokyo

Authors

Desineni S. Naidu
Guidance and Control Division, NASA Langley Research Center
Hampton, VA 23665, USA

Ayalasomayajula K. Rao
Department of Electrical Engineering, J.N. Technical University
Hyderabad, AP. 500028, India

Mathematics Subject Classification (1980): 34 E 15, 39 A 10, 93 C 55

ISBN 3-540-15981-9 Springer-Verlag Berlin Heidelberg New York Tokyo
ISBN 0-387-15981-9 Springer-Verlag New York Heidelberg Berlin Tokyo

Printing and binding: Beltz Offsetdruck, Hemsbach/Bergstr.
2146/3140-543210

Dedicated to

Mother and Father (DSN)

and

Mother (AKR)

ACKNOWLEDGEMENTS

The authors would like to express their deep gratitude to Professor P. K. Rajagopalan of Indian Institute of Technology Kharagpur for introducing the subject to the authors and for his inspiration. One of the authors (DSN) had a gainful experience under the influence of the works of Professor P. V. Kokotovic of University of Illinois, Ubana-Champaign, Professor P. Sannuti of Rutgers University, Piscataway, and Professor R. E. O'Malley, Jr., of Rennselaer Poly-technic Institute, Troy.

The permission given by Taylor & Francis Ltd. (International Journal of control) and John Wiley & Sons (Optimal control: Applications & Methods) to reproduce some of the material published by the authors is acknowledged.

Thanks are due to National Research council (NRC) which offered a Senior Research Associateship to one of the authors (DSN) tenable at Flight Dynamics and Control Division, NASA Langley Research Center, Hampton, where the monograph took the final shape. Special mention should be made about Dr. Douglas B. Price who took keen interest and the entire staff who did a marvellous job in the final preparation of the monograph.

Finally, the authors wish to record their high appreciation of their families who made several cheerful sacrifices during the entire period of their research work.

PREFACE

The dynamics of many control systems is described by higher order differential equations. However, the behaviour is governed by a few dominant parameters, a relatively minor role being played by the remaining parameters such as small time constants, masses, moments of inertia, inductances and capacitances. The presence of these "parasitic" parameters is often the source for the increased order and the "stiffness" of the system. The "curse" of this dimensionality coupled with stiffness causes formidable computational difficulties for the analysis and control of such large systems. The singular perturbation method using the reduced order models and relieving the stiffness, is a "gift" to control engineers. As such it is very attractive to formulate many control problems to fit into the framework of the mathematical theory of singular perturbations which has a rich literature [1-5]. The singular perturbation theory in continuous control systems has reached a certain level of maturity and is well documented [6-11].

Discrete systems are very much prevalent in science and engineering. There are three important sources of discrete models described by high order difference equations containing several small parameters [12]. The first source is digital simulation, where the ordinary differential equations are approximated by the corresponding difference equations [13]. The study of sampled-data control systems and computer-based adaptive control systems leads in a natural way to another source of discrete-time models [14,15]. Finally, many economic, biological and sociological systems are represented by discrete models [16,17]. In spite of its paramount importance, the area of singular perturbations in difference equations and its applications to discrete control problems has not so far received sufficient attention [18-28]. It is in this context that the present investigation is taken up in the field of singular perturbation analysis of discrete control systems. The motivation for the present investigation comes mainly from the work of Comstock and Hsiao [18]. The Monograph starts with the analysis of singularly perturbed difference equations in classical form and the various state space models [29-32] and contributes towards the development of singular perturbation methods for open-loop control [33], and closed-loop optimal control [34]. Various implications in casting the equations in a form suitable for singular perturbation analysis and many distinguishing features of the analysis are examined with relevance to each problem. Typical numerical examples are provided to illustrate the proposed methods.

The Monograph is organized as follows:

In Chapter 1, the singularly perturbed difference equations in classical form are formulated as initial value problems and boundary value problems. Methods are developed to obtain approximate solutions in terms of an "outer series" based on the degenerate (reduced) problem and a "boundary layer correction series" obtained by using certain transformations on the original problem.

In Chapter 2, we consider the state space modelling and analysis of singularly perturbed difference equations in order to give a general framework suitable for control engineers. Depending on the position of the small parameter, three state space discrete models are formulated and techniques are developed to obtain approximate series solutions [29-32]. The computational requirements of singularly perturbed differential equations are tremendous due to their stiffness [35-40]. A method is therefore suggested to cast singularly perturbed differential equations into the corresponding discrete models. The case of sampled-data control systems is also examined.

In Chapter 3, the three-time scale property of difference equations is first examined. Then, the open-loop optimal control of singularly perturbed discrete system is investigated [41]. For the resulting two-point boundary value problem, a method is developed consisting of an outer series and two correction series corresponding to "initial" and "final" boundary layers [33].

In Chapter 4, we investigate the basic ideas of the singularly perturbed nonlinear difference equations. Then the closed-loop optimal control of a linear singularly perturbed discrete system is examined [41]. For solving the resulting matrix Riccati difference equation, a method is developed. The steady state solution of the Riccati equation is also examined [34].

There are several other contributions made in singular perturbation and time scale analysis of discrete systems with reference to general control problems [46-49, 51, 53-55, 57-60, 62, 68-72, 75] optimal control problems [52, 65-67, 73], adaptive control problems [56, 63, 64], and stochastic systems and Markov chains [50, 61, 74].

D. S. Naidu

A. K. Rao

CONTENTS

SINGULAR PERTURBATION ANALYSIS

OF

DIFFERENCE EQUATIONS IN CLASSICAL FORM

In this Chapter the analysis of singularly perturbed difference equations in classical form is carried out. In order to get a clear insight into the analysis, both initial and boundary value problems are first considered in their simplest form. The various distinguishing features of singular perturbations such as order reduction, loss of boundary conditions, boundary layer and boundary layer correction are clearly brought out. Methods are developed where the approximate solution is sought in the form of an outer series solution and a correction series solution. A special feature of this chapter is to give a complete analysis of a general higher order difference equation with several small parameters [29]. Examples are provided to illustrate the proposed methods.

1.1 Small Parameter at the Right End: R-type

As a first step leading to the analysis of higher order difference equations and in order to get a clear insight into the method, consider a second order linear homogeneous equation

$$y(k+2) + ay(k+1) + hy(k) = 0 \qquad \qquad \dots \quad (1.1)$$

where h is a small parameter negligibly small in comparison with the other coefficients a and 1.

The solution of the above equation is

$$y(k) = c_1(m_1)^k + c_2 (m_2)^k$$

where m_1 and m_2 are the characteristic roots of (1.1) given by

$$m_1 = -\frac{a}{2} + \frac{a}{2} (1 - 4h/a^2)^{0.5} \qquad \qquad \dots \quad (1.2a)$$

$$m_2 = -\frac{a}{2} - \frac{a}{2} (1 - 4h/a^2)^{0.5} \qquad \qquad \dots \quad (1.2b)$$

and c_1 and c_2 are constants that depend on the give conditions.

Case (a): Initial Value Problem (IVP)

Given the initial conditions $y(0)$ and $y(1)$, th is

$$y(k) = \frac{(y(0)m_2 - y(1))(m_1)^k + (y(1) - y(0)m_1)}{m_2 - m_1}$$

where m_1 and m_2 are as in (1.2).

Using the Binomial expansion

$$(1 - 4h/a^2)^{0.5} = 1 - 2h/a^2 - 2h^2/a^4 \ldots \ldots$$

if $|4h/a^2| < 1$,

for sufficiently small values of h, the solution power series in h.

Ignoring terms with coefficients of h and hig zeroth order approximate solution is

$$y(k) = y(1)(-a)^{k-1} + h^k [(y(0 + y(1)/a)(-1/a)$$

Similarly, ignoring terms with coefficients powers of h, the first order approximate solution

$$y(k) = y(1)(-a)^{k-1} + h(N\frac{y(0)}{a} + \frac{y(1)}{a^2} + \frac{(1-k)}{a^2} y($$

$$+h^k((y(0)+y(1)/a)(-1/a)^k + hNm(y(0)+y(1)/a)\frac{k}{a^2} +$$

Solutions of higher order approximation can lines.

SINGULAR PERTURBATION ANALYSIS

OF

DIFFERENCE EQUATIONS IN CLASSICAL FORM

In this Chapter the analysis of singularly perturbed difference equations in classical form is carried out. In order to get a clear insight into the analysis, both initial and boundary value problems are first considered in their simplest form. The various distinguishing features of singular perturbations such as order reduction, loss of boundary conditions, boundary layer and boundary layer correction are clearly brought out. Methods are developed where the approximate solution is sought in the form of an outer series solution and a correction series solution. A special feature of this chapter is to give a complete analysis of a general higher order difference equation with several small parameters [29]. Examples are provided to illustrate the proposed methods.

1.1 Small Parameter at the Right End: R-type

As a first step leading to the analysis of higher order difference equations and in order to get a clear insight into the method, consider a second order linear homogeneous equation

$$y(k+2) + ay(k+1) + hy(k) = 0 \qquad \qquad \ldots \quad (1.1)$$

where h is a small parameter negligibly small in comparison with the other coefficients a and 1.

The solution of the above equation is

$$y(k) = c_1(m_1)^k + c_2 (m_2)^k$$

where m_1 and m_2 are the characteristic roots of (1.1) given by

$$m_1 = - \frac{a}{2} + \frac{a}{2} (1 - 4h/a^2)^{0.5} \qquad \qquad \ldots \quad (1.2a)$$

$$m_2 = - \frac{a}{2} - \frac{a}{2} (1 - 4h/a^2)^{0.5} \qquad \qquad \ldots \quad (1.2b)$$

and c_1 and c_2 are constants that depend on the given boundary conditions.

Case (a): Initial Value Problem (IVP)

Given the initial conditions $y(0)$ and $y(1)$, the solution of (1.1) is

$$y(k) = \frac{(y(0)m_2 - y(1))(m_1)^k + (y(1) - y(0)m_1)\ (m_2{}^k)}{m_2 - m_1} \qquad \dots \quad (1.3)$$

where m_1 and m_2 are as in (1.2).

Using the Binomial expansion

$$(1 - 4h/a^2)^{0.5} = 1 - 2h/a^2 - 2h^2/a^4 \ . \ . \ . \ . \ . \ .$$

if $|4h/a^2| < 1$,

for sufficiently small values of h, the solution (1.3) is obtained as a power series in h.

Ignoring terms with coefficients of h and higher powers of h, the zeroth order approximate solution is

$$y(k) = y(1)(-a)^{k-1} + h^k\ [(y(0 + y(1)/a)(-1/a)^k] \qquad \dots \quad (1.4a)$$

Similarly, ignoring terms with coefficients of h^2 and higher powers of h, the first order approximate solution is

$$y(k) = y(1)(-a)^{k-1} + h(N\frac{y(0)}{a} + \frac{y(1)}{a^2} + \frac{(1-k)}{a^2}\ y(1)0(-a)^{k-1}|$$

$$+h^k((y(0)+y(1)/a)(-1/a)^k+hNm(y(0)+y(1)/a)\frac{k}{a^2} + \frac{y(0)}{a^2}\ \frac{2y(1)}{a^3}\%m\ -1/a\%^k)|$$

$$\dots \quad (1.4b)$$

Solutions of higher order approximation can be obtained on similar lines.

Note that in (1.4), h^k is included to take account of the fact that the contribution to the series associated by this term is different when h and k tend to zero in different sequences (ie. h T 0 first and k T 0 later or vice versa).

Suppressing the small parameter h in (1.1), the resulting degenerate equation is

$$y^{(0)}(k+2) + a\, y^{(0)}(k+1) = 0 \qquad \ldots \quad (1.5)$$

This equation is of order one and can satisfy only one of the given two initial conditions. Hence (1.1) is said to be in the singularly perturbed form [18].

If (1.5) is solved with $y^{(0)}(1) = y(1)$, the solution obtained is

$$y^{(0)}(k) = y(1)\,(-a)^{k-1} \qquad \ldots \quad (1.6)$$

which is the same as the first term in (1.4a).

The above solution yields

$$y^{(0)}(0) = -y(1)/a \quad] \quad y(0).$$

A close examination of (1.4) reveals that

(i) Limit y(k)   ]   Limit y(k)
       h T 0   k T 0
       k T 0   h T 0

In other words, as h tends to zero the uniform convergence of the solution (1.3) to the degenerate solution (1.6) fails at k = 0. The degenerate solution is therefore not valid at that point and a <u>boundary layer occurs at k = 0.</u>

(ii) The solution in (1.4) not involving h^k is called the "<u>outer series</u>" solution, as it is valid outside the boundary layer. This satisfies only one of the given initial conditions y(1). The solution involving h^k is called the "<u>boundary layer correction series</u>" solution which recovers the lost boundary condition y(0).

(iii) The presence of h^k in the series solution (1.4) suggests that the correction series solution has the transformation

$$w(k) = y(k)/h^k$$

Using the above transformation in (1.1) and dividing throughout with h^{k+1}, the equation of the correction series is obtained as

$$hw(k+2) + aw(k+1) + w(k) = 0 \qquad \qquad \dots \quad (1.7)$$

Putting $h = 0$ in the above equation, its degenerate form is

$$w^{(0)}(k+1) + \frac{1}{a} w^{(0)}(k) = 0$$

The above equation is now solved with the initial condition

$$w^{(0)}(0) = y(0) - y^{(0)}(0)$$

Yielding

$$w^{(0)}(k) = (y(0) - y^{(0)}(0)) \ (-1/a)^k$$

Using the value of $y^{(0)}(0)$ from (1.6),

$$w^{(0)}(k) = (y(0) + y(1)/a) \ (-1/a)^k$$

Note that the above solution is the same as the term associated with h^k in (1.4a). This illustrates the recovery of the initial condition lost in the process of degeneration.

Thus the total zeroth order series solution composed of the outer series solution and the correction series solution is given by

$$y(k) = y^{(0)}(k) + h^k w^{(0)}(k)$$

The above solution, not only satisfies both the given initial conditions but also is nearer to the exact solution (1.3) upto zeroth order approximation.

Development of the singular perturbation method

For a detailed analysis, utilizing the previous results, the solution $y(k)$ is written as the sum of two solutions,

$$y(k) = y_t(k) + h^k w(k) \qquad \dots \quad (1.8)$$

where $y_t(k)$ and $w(k)$ are called the solutions of outer series and the boundary layer correction series respectively. Substituting (1.8) in (1.1) and separating the terms of outer series and correction series, two separate equations are obtained as below.

$$y_t(k+2) + ay_t(k+1) + hy_t(k) = 0 \qquad \dots \quad (1.9a)$$

$$hw(k+2) + aw(k+1) + w(k) = 0 \qquad \dots \quad (1.9b)$$

In the above two equations, the first one corresponds to the outer series and the second one to the correction series. It is to be noted that the first equation is nothing but the original equation (1.1). For solving the equations (1.9), series solutions are assumed as

$$y_t(k) = \sum_{r=0}^{\infty} h^r y^{(r)}(k); \quad w(k) = \sum_{r=0}^{\infty} h^r w^{(r)}(k) \qquad \dots \quad (1.10)$$

Substituting the above series solutions (1.9), the equations for the outer series and correction series for various orders of approximation are obtained as below.

Outer Series:

$$h^0 : y^{(0)}(k+2) + ay^{(0)}(k+1) = 0$$

$$h^1 : y^{(1)}(k+2) + ay^{(1)}(k+1) = -y^{(0)}(k)$$

$$\dots \quad \dots \dots \dots \dots \dots \dots \dots$$

$$h^j : y^{(j)}(k+2) + ay^{(j)}(k+1) = -y^{(j-1)}(k) \qquad \dots \quad (1.11a)$$

Correction Series:

h^0 : $a \, w^{(0)}(k+1) + w^{(0)}(k) = 0$

h^1 : $a \, w^{(1)}(k+1) + w^{(1)}(k) = -w^{(0)}(k+2)$

... .

h^j: $a \, w^{(j)}(k+1) + w^{(j)}(k) = -w^{(j-1)}(k+2)$ ... (1.11b)

Adding the outer series and the correction series terms, the total series solution is written as

$y(k) = y^{(0)}(k) + h y^{(1)}(k) + h^2 y^{(2)}(k) + . . . \cdot . .$

$+ h^k [w^{(0)}(k) + h w^{(1)}(k) + h^2 w^{(2)}(k) + ...]$

$= \displaystyle\sum_{r=0}^{j} [y^{(r)}(k) + h^k w^{(r)}(k)] \, h^r$ ... (1.12)

The initial conditions required to solve the above series equations are obtained from the total series solution (1.12) based on the fact that the two series together satisfy all the given initial conditions exactly. They are furnished as shown below.

h^0 : $y^{(0)}(1) = y(1)$; $w^{(0)}(0) = y(0) - y^{(0)}(0)$

h^1 : $y^{(1)}(1) = - w^{(0)}(1)$; $w^{(1)}(0) = - y^{(1)}(0)$

... .

h^j : $y^{(j)}(1) = - w^{(j-1)}(1)$; $w^{(j)}(0) = - y^{(j)}(0)$

From (1.12), it follows that it is enough if the correction series equations (1.11b) are solved for a few values of k near the boundary layer at k = 0. This is due to h^k multiplying the correction series terms. Suppose we are interested in the solution up to second order approximation. Then it is apparent from (1.12) that for zeroth order solution, only $w^{(0)}(0)$ is required, for first order solution, only $w^{(0)}(1)$ and $w^{(1)}(0)$ are required additionally, and for second order solution, only $w^{(0)}(2)$, $w^{(1)}(1)$ and $w^{(2)}(0)$ are required additionally.

Example 1.1

Consider a second order difference equation

$$y(k+2) - 0.8y(k+1) - hy(k) = 0$$

with $y(0) = 2.0$, $y(1) = 1.0$ as the given initial conditions and $h = 0.08$. The two characteristic roots of the above equation are

$$m_1 = 0.8899, \quad m_2 = -0.0899.$$

Fig. 1.1 illustrates how the solution $y(k)$ approaches the degenerate solution as h tends to zero and the occurrence of the boundary layer at $k = 0$.

Using the method developed, the results are obtained for degenerate, zeroth, first and second order solutions and compared with the exact solution as shown in Table 1.1 and Fig. 1.2.

The results given in Table 1.1 and Fig. 1.2 clearly indicate that as the order of approximation increases, the series solutions approach the exact solution.

Note: (i) From the results of Table 1.1, it is noted that the first order and exact solutions coincide for $k = 0$ to 3. Similarly the second order and exact solutions coincide for $k = 0$ to 5. This is due to the fact that if the solution of the above equation is found upto $k = 3$, no terms with h^2 and higher powers of h appear. Similarly, upto $k = 5$, no terms with h^3 and higher powers of h appear. Thus for the initial value problem, the solutions for various orders of approximation are the same as the exact solution upto a few values of k.

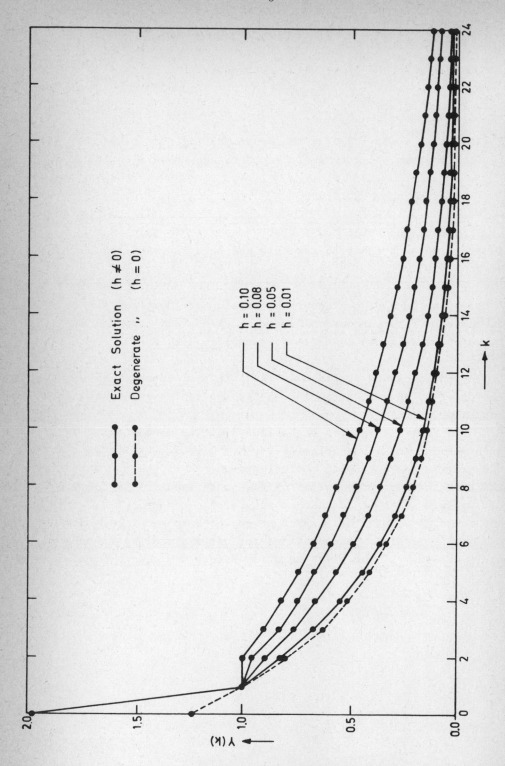

Fig. 1.1 Illustration of degeneration of the solution of Example 1.1

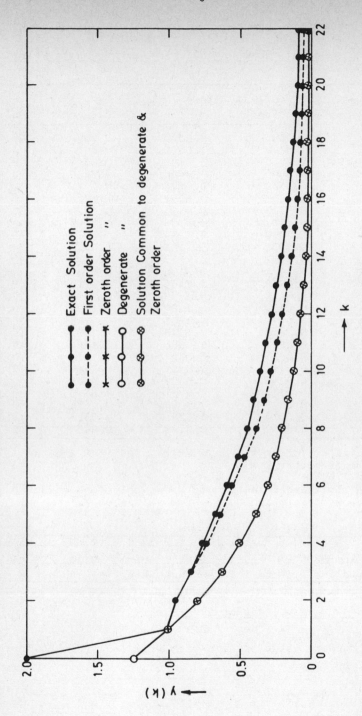

Fig. 1.2 Exact and approximate solutions of Example 1.1

Table 1.1

Comparison of approximate solutions with the exact solution
of Example 1.1

k	Degenerate solution	Zeroth order solution	1st order solution	2nd order solution	Exact solution
0	1.2500	2.0000	2.0000	2.0000	2.0000
1	1.0000	1.0000	1.0000	1.0000	1.0000
2	0.8000	0.8000	0.9600	0.9600	0.9600
3	0.6400	0.6400	0.8480	0.8480	0.8480
4	0.5120	0.5120	0.7424	0.7552	0.7552
5	0.4096	0.4096	0.6451	0.6720	0.6720
6	0.3277	0.3277	0.5571	0.5970	0.5980
7	0.2621	0.2621	0.4784	0.5292	0.5322
8	0.2097	0.2097	0.4089	0.4679	0.4736
9	0.1678	0.1678	0.3481	0.4126	0.4214
10	0.1342	0.1342	0.2953	0.3628	0.3750
11	0.1074	0.1074	0.2496	0.3181	0.3337
12	0.0859	0.0859	0.2105	0.2781	0.2970
13	0.0687	0.0687	0.1770	0.2425	0.2643
14	0.0550	0.0550	0.1484	0.2108	0.2352
15	0.0440	0.0440	0.1242	0.1828	0.2093
16	0.0352	0.0352	0.1038	0.1581	0.1863
17	0.0281	0.0281	0.0866	0.1364	0.1658
18	0.0225	0.0225	0.0721	0.1174	0.1475
19	0.0180	0.0180	0.0599	0.1009	0.1313
20	0.0144	0.0144	0.0497	0.0865	0.1168
21	0.0115	0.0115	0.0412	0.0740	0.1039
22	0.0092	0.0092	0.0341	0.0632	0.0925
23	0.0074	0.0074	0.0282	0.0538	0.0823
24	0.0059	0.0059	0.0233	0.0458	0.0733

Note: 1) The zeroth order and degenerate solutions are the
same except at k = 0.

2) The first order and second order solutions coincide with
the exact solution upto a few values of k.

(ii) In Figs. 1.1 and 1.2, the solutions obtained at discrete
values of k are joined by straight lines, for the sake of clarity.
This is followed for all the figures throughout the monograph.

Case (b): Boundary value problem (BVP)

For the boundary value problem with y(0) and y(N) known, the solution of (1.1) is

$$y(k) = \frac{(y(0)m_2{}^N - y(N))(m_1)^k + (y(N)-y(0)m_1{}^N)(m_2)^k}{m_2{}^N - m_1{}^N} \qquad \ldots \quad (1.13)$$

where m_1 and m_2 are as in (1.2).

The zeroth order approxmite solution is obtained from (1.13) as

$$y(k) = y(N)(-a)^{k-N} + h^k \left[\{y(0)-y(N)(-a)^{-N}\}\{-1/a\}^k\right] \qquad \dots \quad (1.14a)$$

The first order approximate solution is

$$y(k) = y(N)(-a)^{k-N} + h\left[\left\{\frac{(N-K)}{a^2} \, y(N)\right\} \{-a\}^{k-N}\right]$$

$$+ h^k \left[\{y(0)-y(N)(-a)^{-N}\}\{-1/a\}^k\right.$$

$$+ h\left\{\frac{ky(0)}{a^2} - \frac{(k+N)y(N)}{a^2}\right\} (-1/a)^{k-N}\right]$$

$$\dots \quad (1.14b)$$

A close examination of (1.14) reveals that as in IVP,

$$\begin{array}{ccc} \text{Limit } y(k) & \neq & \text{Limit } y(k) \\ h \to 0 & & k \to 0 \\ k \to 0 & & h \to 0 \end{array}$$

In other words, as h tends to zero the exact solution fails to uniformly converge to the degenerate solution at k = 0. This leads to the formation of the boundary layer at k = 0 and in the process of degeneration, the initial value y(0) is lost. Further, (1.14) suggests that the boundary layer correctiion series is obtained as in IVP, using the same transformation

$$w(k) = y(k)/h^k$$

Thus the equation for the correction series is the same as (1.9b). Proceeding on similar lines as in IVP, series equations for various orders of approximation for the outer series and correction series are obtained, which are same as (1.11a) and (1.11b) respectively. The selection of the boundary values for solving these equations is however different. They are obtained from the total series solution (1.12) and furnished as shown.

$h^0 : y^{(0)}(N) = y(N); w^{(0)}(0) = y(0) - y^{(0)}(0)$

$h^1 : y^{(1)}(N) = 0 \quad ; w^{(1)}(0) = - y^{(1)}(0)$

$\cdots \cdots \cdots \cdots \cdots \cdots \cdots \cdots \cdots$

$h^j : y^{(j)}(N) = 0 \quad ; w^{(j)}(0) = -y^{(j)}(0)$

1.2 Small Parameter at the Left End: L-type [18]

Consider a different class of equations where the singular perturbation parameter h is associated with <u>left</u> end as shown

$$hy(k+2) + ay(k+1) + y(k) = 0 \qquad \cdots \quad (1.16)$$

The solution of (1.16) is written as

$$y(k) = c_1(m_1)^k + c_2(m_2)^k$$

where m_1 and m_2 are the characteristic roots of (1.16) given by

$$m_1 = - \frac{a}{2h} + \frac{a}{2h} (1 - 4h/a^2)^{0.5} \qquad \cdots \quad (1.17a)$$

$$m_2 = - \frac{a}{2h} - \frac{a}{2h} (1 - 4h/a^2)^{0.5} \qquad \cdots \quad (1.17b)$$

and c_1 and c_2 are constants that depend on the given boundary
conditions.

Given $y(0)$ and $y(N)$, the solution of (1.16) is

$$y(k) = \frac{[y(0)m_2^N - y(N)](m_1)^k + [y(N) - y(0)m_1^N](m_2)^k}{m_2^N - m_1^N} \qquad \cdots \quad (1.18)$$

where m_1 and m_2 are as in (1.17).

Using the Binomial expansion

$$(1-4h/a^2)^{0.5} = 1 - \frac{2h}{a^2} - \frac{2h^2}{a^4} + \cdots , \text{ if } \left|\frac{4h}{a^2}\right| < 1,$$

given y(0) and y(1), the zeroth order approximate solution for the initial value problem is

$$y(k) = y(0)(-1/a)^k + h^{1-k}[(y(1)-y(0)(-1/a))(-a)^{k-1}]$$

In the above equation for $k > 2$, it is seen that the solution of y(k) cannot be expressed in positive powers of h. In addition, the solution of (1.16) is always <u>unstable.</u> Therefore a singular perturbation method for IVP <u>cannot</u> be developed for this class of equations.

However for BVP where y(0) and y(N) are the given boundary conditions, the method can be developed. Ignoring terms with coefficients of h and higher powers of h, the zeroth order approximate solution is obtained from (1.18) as

$$y(k) = y(0)(-1/a)^k + h^{N-k}[(y(N)-y(0)(-1/a)^N)(-a)^{k-N}]$$

... (1.19)

A close examination of (1.19) reveals that

(i) If the degenerate equation of (1.16) is solved with $y^{(0)}(0) = y(0)$, the solution obtained is

$$y^{(0)}(k) = y(0)(-1/a)^k$$

which is the same as the first term in (1.19)

(ii) As h tends to zero, the solution y(k) of (1.18) fails to uniformly converge to the degenerate solution $y^{(0)}(k)$ at $k = N$, thereby indicating the <u>occurrence of the boundary layer at k = N.</u> As a result, the boundary condition y(N) is lost in the process of degeneration.

(iii) The lost boundary condition y(N) can be recovered by the correction series using the transformation

$$w(k) = y(k)/h^{N-k}$$

Using the above transformation in (1.16), the equation of the correction series is obtained as

$$w(k+2) + aw(k+1) + hw(k) = 0$$

The procedure of obtaining the equations for the outer series and correction series is the same as in the previous section. The series equation are furnished below.

Outer Series:

$$h^0 \;:\; a\, y^{(1)}(k+1) + y^{(0)}(k) = 0$$

$$h^1 \;:\; a\, y^{(1)}(k+1) + y^{(1)}(k) = -\, y^{(0)}(k+2)$$

$$\cdots \qquad \cdots \cdots \cdots \cdots \cdots \cdots$$

$$h^j \;:\; a\, y^{(j)}(k+1) + y^{(j)}(k) = -\, y^{(j-1)}(k+2)$$

Correction Series:

$$h^0 \;:\; w^{(0)}(k+2) + a\, w^{(0)}(k+1) = 0$$

$$h^1 \;:\; w^{(1)}(k+2) + a\, w^{(1)}(k+1) = -w^{(0)}(k)$$

$$\cdots \qquad \cdots \cdots \cdots \cdots \cdots \cdots$$

$$h^j \;:\; w^{(j)}(k+2) + a\, w^{(j)}(k+1) = -w^{(j-1)}(k)$$

The total series solution is

$$y(k) = \sum_{r=0}^{j} \left[y^{(r)}(k) + h^{N-K} w^{(r)}(k) \right] h^r$$

The boundary conditions required to solve the above series equations are

$$h^0 \;:\; y^{(0)}(0) = y(0); \; w^{(0)}(N) = y(N) - y^{(0)}(N)$$

$$h^1 \;:\; y^{(1)}(0) = 0 \quad ; \; w^{(1)}(N) = -\, y^{(1)}(N)$$

$$\cdots \qquad \cdots \cdots \cdots \cdots \cdots \cdots \cdots$$

$$h^j \;:\; y^{(j)}(0) = 0 \quad ; \; w^{(j)}(N) = -\, y^{(j)}(N)$$

Note: It is noted that an interrelationship exists between the R- and L-type equations as explained below.

Consider the R-type equation

y(k+2) + ay(k+1) + hy(k) = 0

Repacing k by N-k, the above equation is written as

y(N-k-2) + ay(N-k-1) + hy(N-k) = 0

or

hy(N-k+2) + ay(N-k+1) + y(N-k) = 0

The above equation is evidently of L-type. On similar lines it is possible to get the R-type equation from the L-type.

It is interesting to note that if m_1 and m_2 are the characteristic roots of the R-type equation, the corresponding roots of the L-type equation obtained as above are $\frac{1}{m_2}$ and $\frac{1}{m_1}$ respectively. This is seen from (1.2) and (1.17).

Example 1.2

Consider the boundary value problem

hy(k+2) - 0.9y(k+1) + y(k) = 0

y(0) = 1.0 and y(0) = 10.0 are the given boundary conditions and h = 0.08.

The characteristic roots of the above equation are

m_1 = 10.0, m_2 = 1.25

Fig. 1.3 illustrates how the solution y(k) approaches the degenerate solution as h tends to zero, and the formation of the boundary layer at k = 10 (the terminal point).

Using the method developed above, the results are obtained for degenerate, zeroth, first and second order solutions and compared with the exact solution in Table 1.2 and Fig. 1.4. As the order of approximation increases, the results indicate that the series solutions approach the exact solution.

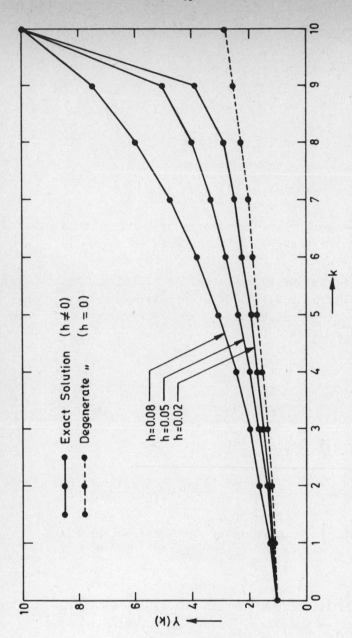

Fig. 1.3 Illustration of degeneration of the solution of Example 1.2

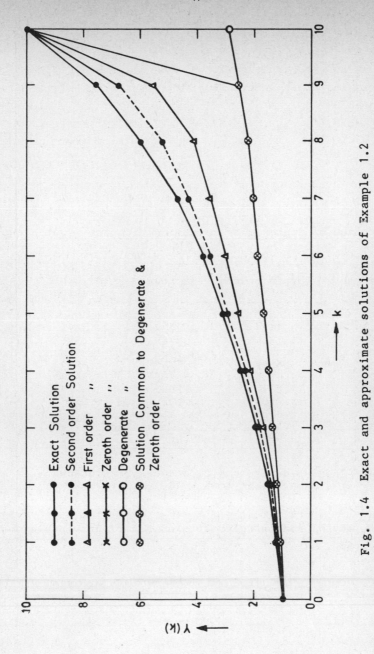

Fig. 1.4 Exact and approximate solutions of Example 1.2

1.3 Higher Order Difference Equations [29]

Consider an (n+m)th order linear, constant coefficient difference equation with small parameters occurring at the right end as

$$c_{m+n} \, y(k+m+n) + c_{m+n-1} \, y(k+m+n-1) \; \ldots \; + c_m y(k+m)$$

$$+ \; c_{m-1} \, y(k+m-1) + \ldots + c_1 y(k+1) + c_0 y(k) = b_0 u(k) \; \ldots (1.20)$$

Table 1.2

Comparison of approximate solutions with the exact solution
of Example 1.2

k	Degenerate solution	Zeroth order solution	1st order solution	2nd order solution	Exact solution
0	1.0000	1.0000	1.0000	1.0000	1.0000
1	1.1111	1.1111	1.2209	1.2425	1.2500
2	1.2346	1.2346	1.4784	1.5386	1.5625
3	1.3717	1.3717	1.7782	1.8986	1.9531
4	1.5242	1.5242	2.1263	2.3344	2.4414
5	1.6935	1.6935	2.5298	2.8602	3.0518
6	1.8817	1.8817	2.9967	3.4923	3.8148
7	2.0908	2.0908	3.5362	4.2500	4.7691
8	2.3231	2.3231	4.1586	5.2120	5.9612
9	2.5812	2.5812	5.5095	6.6000	7.5193
10	2.8680	10.0000	10.0000	10.0000	10.0000

Note: The degenerate and zeroth order solutions are the same except at k = 10.

where the m coefficients c_0, c_1, $\ldots$ c_{m-1} are negligibly small in comparison with the remaining coefficients. By a suitable choice of coefficients as

$$\frac{c_{m+n-1}}{c_{m+n}} = a_{m+n-1}, \qquad \frac{c_{m+n-2}}{c_{m+n}} = a_{m+n-2}, \; \cdots \; \frac{c_m}{c_{m+n}} = a_m, \; \text{and}$$

$$\frac{c_{m-1}}{c_{m+n}} = a_{m-1} \, h, \qquad \frac{c_{m-2}}{c_{m+n}} = a_{m-2} h^2, \; \cdots \; \frac{c_1}{c_{m+n}} = a_1 h^{m-1},$$

$$\frac{c_0}{c_{m+n}} = a_0 h^m \; \text{and} \; b_0 = \frac{d_0}{c_{m+n}},$$

where h is a small positive parameter, much smaller than one, the above equation can be expressed in the general form of R-type as

$$y(k+m+n) + a_{m+n-1}y(k+m+n-1) + \ldots + a_m y(k+m)$$

$$+ a_{m-1} h\, y(k+m-1)+\ldots\ldots+a_1 h^{m-1}y(k+1)+a_o h^m y(k) = b_o u(k)\ldots \quad (1.21)$$

The above equation is in the singularly perturbed form in the sense that by making $h = 0$ in it, the order of the equation drops from $(n+m)$ to n. That is, the reduced or degenerate equation becomes

$$y^{(0)}(k+m+n) + a_{m+n-1}y^{(0)}(k+m+n-1)+ \ldots+a_m y^{(0)}(k+m)=b_o u(k).$$

The external input $u(k)$ is assumed to be independent of h.

Case (a): Initial value problem

To examine the nature of the solution of a difference equation containing several small parameters for the sake of simplicity a second-order homogeneous equation is considered as

$$y(k+2) + ahy(k+1) + bh^2 y(k) = 0 \qquad \qquad \ldots \quad (1.22)$$

where h is a small positive parameter.

The two roots of the auxiliary equation of (1.22) are given by

$$m_1,m_2 = h\ [- \frac{a}{2} \pm \frac{a}{2}\ (1-4b/a^2)^{0.5}\ ] = a_1 h, a_2 h$$

In terms of the given intial conditions $y(0)$ and $y(1)$, the solution of (1.22) is written as

$$y(k) = h^{k-1}[\ \frac{y(1)}{(a_1-a_2)}\ (a_1^k - a_2^k)] + h^k[\frac{y(0)}{(a_1-a_2)}\ (a_1 a_2^k - a_2 a_1^k)]$$

$$\ldots \quad (1.23)$$

From (1.22), it is evident that its reduced or degenerate equation has a trivial solution. From (1.23), as $h \to 0$, $y(k) \to 0$ except at $k = 1$ and $k = 0$. Therefore in the process of degeneration, the initial conditions are lost at $k = 1$ and $k = 0$. The nature of the exact solution suggests that one can recover the lost initial conditions by incorporating two separate correction series equations obtained by transforming (1.22), with

$$w_0(k) = y(k)/h^{k-1}; \quad w_0(k) = y(k)/h^k \qquad \qquad \cdots \quad (1.24)$$

Using the above transformations in (1.22) the equations of the two correction series are

$$w_1(k+2) + aw_1(k+1) + bw_1(k) = 0$$

$$w_0(k+2) + aw_0(k+1) + bw_0(k) = 0$$

The above equations do not contain the terms of h as a special case, since the reduced equation of (1.22) is of zero order. The reduced form of these correction series equations is

$$w_1^{(0)}(k+2) + aw_1^{(0)}(k+1) + bw_1^{(0)}(k) = 0$$

$$w_0^{(0)}(k+2) + aw_0^{(0)}(k+1) + bw_0^{(0)}(k) = 0$$

Choosing $w_1^{(0)}(0) = 0$, $w_1^{(0)}(1) = y(1)$, $w_0^{(0)}(0) = y(0)$, $w_0^{(0)}(1) = 0$ the solutions of the above equations are

$$w_1^{(0)}(k) = \frac{y(1)}{(a_1 - a_2)} (a_1^k - a_2^k) \qquad \qquad \cdots \quad (1.25a)$$

$$w_0^{(0)}(k) = \frac{y(0)}{(a_1 - a_2)} (a_1 a_2^{\;k} - a_2 a_1^{\;k}) \qquad \qquad \cdots \quad (1.25b)$$

Using the transformations (1.24a, b) when $w_1^{(0)}(k)$ and $w_0^{(0)}(k)$ are added to the trivial degenerate solution, we get the total series solution which is the same as (1.23). The solution thus becomes

$$y(k) = h^{k-1} w_1^{(0)}(k) + h^k w_0^{(0)}(k) ,$$

where $w_1(k)$ and $w_0(k)$ are the two separate correction series for the recovery of y(1) and y(0) respectively.

The method of recovery of the lost initial conditions is thus clearly demonstrated in the case of a simple second-order system.

Based on the above analysis, the solution of the $(n + m)$th order difference equation (1.21) is assumed as the sum of the outer series and m number of correction series as given below.

$$y(k) = \sum_{j=o}^{\infty} h^j y_t^{(j)}(k) + h^{k-(m-1)} \sum_{j=o}^{\infty} h^j w_{m-1}^{(j)}(k) + h^{k-(m-2)} \sum_{j=o}^{\infty} h^j w_{m-2}^{(j)}(k)$$

$$+ \ldots\ldots\ldots + h^k \sum_{j=o}^{\infty} h^j w_o^{(j)}(k) \qquad \qquad \ldots \quad (1.26)$$

where $\sum_{j=o}^{\infty} y_t^{(j)}(k)$ is the outer series, and

$$\sum_{j=o}^{\infty} w_s^{(j)}(k), \quad s = m-1, \ m-2, \ \ldots\ldots 0$$

is the correction series to recover the initial condition $y(s)$.

For zeroth order approximation letting $j = 0$, (1.26) becomes

$$y(k) = y_t^{(0)}(k) + h^{k-(m-1)} w_{m-1}^{(0)}(k) + h^{k-(m-2)} w_{m-2}^{(0)}(k) + \ldots$$

$$\ldots\ldots + h^k w_0^{(0)}(k)$$

From the above, it is clear that as h tends to zero, $y(k)$ uniformly coverges to $y_t^{(0)}(k)$ except at $k = 0,1,\ldots\ldots m-1$ and we say m boundary layers occur at these points. The given initial conditions at these values of k are therefore lost in the process of degeneration. Based on the analysis of the second order system, the transformation required to obtain the correction series for the recovery of these m initial conditions are

$$w_{m-1}(k) = y(k)/h^{k-(m-1)}, \quad w_{m-2}(k) = y(k)/h^{k-(m-2)}, \quad \ldots$$

$$w_0^{(0)}(k) = y(k)/h^k \qquad \ldots \quad (1.27a)$$

Denoting

$$y_t(k) = \sum_{j=o}^{\infty} h^j y_t^{(j)}(k)$$

and

$$y_s(k) = h^{k-s} \sum_{j=o}^{\infty} h^j w_s^{(j)}(k), \quad s = m-1, m-2, \ldots \quad 0 \quad \ldots \quad (1.27b)$$

(1.26) can be written as

$$y(k) = y_t(k) + y_{m-1}(k) + y_{m-2}(k) + \ldots + y_o(k)$$

Substituting the above expression for y(k) in the original difference equation (1.21), and separating the outer series and the correction series terms the following equations are obtained.

$$y_t(k+m+n) + a_{m+n-1} y_t(k+m+n-1) + \ldots + a_{m-1} h y_t(k+m-1)$$

$$\ldots + a_1 h^{m-1} y_t(k+1) + a_o h^m y_t(k) = b_o u(k) \qquad \ldots \quad (1.28)$$

$$y_s(k+m+n) + a_{m+n-1} y_s(k+m+n-1) + \ldots + a_{m-1} h y_s(k+m-1)$$

$$\ldots + a_1 h^{m-1} y_s(k+1) + a_o h^m y_s(k) = 0, \quad s = m-1, m-2 \ldots 0$$
$$\ldots \quad (1.29)$$

In the above two equations (1.28) is the equation for the outer
series, which is the same as the original difference equation (1.21).
Using the transformations (1.27a) in (1.29), m equations for the correction series are obtained as below.

$$h^{k+n+1} w_s(k+m+n) + a_{m+n-1} h^{k+n} w_s(k+m+n-1) + \ldots$$

$$+ a_{m+1} h^{k+2} w_s(k+m+1) + a_m h^{k+1} w_s(k+m) + \ldots + a_o h^{k+1} w_s(k) = 0$$

$$s = m-1, \; m-2, \; \ldots \ldots \; 0$$

Dividing the above equation throughout with h^{k+1}

$$h^n w_s(k+m+n) + a_{m+n-1} h^{n-1} w_s(k+m+n-1) + \ldots$$

$$+ a_{m+1} h \, w_s(k+m+1) + a_m w_s(k+m) + \ldots + a_o w_s(k) = 0$$

$$s = m - 1, \; m - 2, \; \ldots \ldots \; 0$$

Assuming series expansions as in (1.27b), the equations for zeroth, first and higher order approximations are obtained as shown.

Outer series:

$$h^0 : \; y^{(0)}(k+m+n) + a_{m+n-1} y^{(0)}(k+m+n-1) + \ldots \ldots$$

$$+ a_m \, y^{(0)}(k+m) = b_0 u(k)$$

$$h^1 : \; y^{(1)}(k+m+n) + a_{m+n-1} y^{(1)}(k+m+n-1) + \ldots + a_m y^{(1)}(k+m)$$

$$= - a_{m-1} \, y^{(0)}(k+m-1)$$

$$.. \quad \ldots \ldots \ldots \ldots \ldots \ldots \ldots \ldots \ldots \ldots \ldots \ldots$$

$$h^j : \; y^{(j)}(k+m+n) + a_{m+n-1} y^{(j)}(k+m+n-1) + \ldots + a_m y^{(j)}(k+m)$$

$$= -a_{m-1} \, y^{(j-1)}(k+m-1) - a_{m-2} y^{(j-2)}(k+m-2) - \ldots - a_0 y^{(j-m)}(k)$$

$$\ldots \quad (1.30)$$

Correction series:

For $s = 0,1,\ldots m-1$

$$h^0 : a_m w_s^{(0)}(k+m) + a_{m-1} w_s^{(0)}(k+m-1) + \ldots + a_0 w_s^{(0)}(k) = 0$$

$$h^1 : a_m w_s^{(1)}(k+m) + a_{m-1} w_s^{(1)}(k+m-1) + \ldots + a_0 w_s^{(1)}(k)$$

$$= -a_{m+1} w_s^{(0)}(k+m+1)$$

$$\cdot\cdot \quad \cdot\,\cdot\,\cdot\,\cdot\,\cdot\,\cdot\,\cdot\,\cdot\,\cdot\,\cdot\,\cdot\,\cdot\,\cdot\,\cdot\,\cdot\,\cdot\,\cdot\,\cdot$$

$$h^j : a_m w_s^{(j)}(k+m) + a_{m-1} w_s^{(j)}(k+m-1) + \ldots + a_0 w_s^{(j)}(k)$$

$$= -a_{m+1} w_s^{(j-1)}(k+m+1)$$

$$-a_{m+2} w_s^{(j-2)}(k+m+2) - \ldots - a_{m+j} w_s^{(0)}(k+m+j) \qquad \ldots \quad (1.31)$$

The _boundary conditions_ required to solve the above series equations are readily obtained from the total series solution (1.26) as

Outer Series:

$$h^0 : y_t^{(0)}(m) = y(m), \; y_t^{(0)}(m+1) = y(m+1), \; \ldots\ldots$$

$$y_t^{(0)}(m+n-1) = y(m+n-1)$$

$$h^1 : y_t^{(1)}(m) = -w_{m-1}^{(0)}(m)$$

$$y_t^{(1)}(m+1) = y_t^{(1)}(m+2) = \ldots = y_t^{(1)}(m+n-1) = 0$$

$$h^2 : y_t^{(2)}(m) = -w_{m-1}^{(1)}(m) - w_{m-2}^{(0)}(m); \quad y_t^{(2)}(m+1) = -w_{m-1}^{(0)}(m+1)$$

$$y_t^{(2)}(m+2) = y_t^{(2)}(m+3) = \ldots\ldots = y_t^{(2)}(m+n-1) = 0$$

$\ldots \quad \cdot \cdot \cdot \cdot \cdot \cdot \cdot \cdot \cdot \cdot \cdot \cdot \cdot \cdot \cdot \cdot \cdot \cdot$

$$h^j : y_t^{(j)}(m) = -w_{m-1}^{(j-1)}(m) - w_{m-2}^{(j-2)}(m) \ldots\ldots - w_{m-j}^{(0)}(m)$$

$$y_t^{(j)}(m+1) = -w_{m-1}^{(j-2)}(m+1) - w_{m-2}^{(j-3)}(m+1) - \ldots\ldots$$

$$- w_{m-1}^{(0)}(m+j-1)$$

$\ldots \quad \cdot \cdot \cdot \cdot \cdot \cdot \cdot \cdot \cdot \cdot \cdot \cdot \cdot \cdot \cdot \cdot \cdot \cdot$

$$y_t^{(j)}(m+j-1) = -w_{m-1}^{(0)}(m+j-1)$$

$$y_t^{(j)}(m+j) = y_t^{(j)}(m+j+1) \ldots\ldots = y_t^{(j)}(m+n-1) = 0$$

Correction Series:

For $s = m-1, m-2, \ldots\ldots 0$

$$h^0 : w_s^{(0)}(s) = y(s) - y_t^{(0)}(s)$$

$$w_s^{(0)}(0) = w_s^{(0)}(1) = \ldots\ldots = w_s^{(0)}(s-1) = w_s^{(0)}(s+1)$$

$$\ldots\ldots = w_s^{(0)}(m-1) = 0$$

$$h^1 : w_s^{(1)}(s) = -y_t^{(1)}(s)$$

$$w_s^{(1)}(0) = w_s^{(1)}(1) = \ldots\ldots = w_s^{(1)}(s-1) = w_s^{(1)}(s+1)$$

$$\ldots\ldots = w_s^{(1)}(m-1) = 0$$

$$h^2 : w_s^{(2)}(s) = - y_t^{(2)}(s)$$

$$w_s^{(2)}(0) = w_s^{(2)}(1) = \ldots = w_s^{(2)}(s-1) = w_s^{(2)}(s+1) =$$

$$\ldots = w_s^{(2)}(m-1) = 0$$

.. . .

$$h^j : w_s^{(j)}(s) = - y_t^{(j)}(s)$$

$$w_s^{(j)}(0) = w_s^{(j)}(1) = \ldots = w_s^{(j)}(s-1) = w_s^{(j)}(s+1)$$

$$\ldots = w_s^{(j)}(m-1) = 0$$

Proceeding in a systematic manner, all the sets of equations can be solved with the above initial conditions and the total series solution is evaluated from (1.26) for any desired order of approximation.

It is also important to note that the order of the outer series equations is n, while that for the correction series is m. Further the correction series equations (1.31) need not be solved for all values of , since terms like $h^{k-(m-1)}$, $h^{k-(m-2)}$, etc. appear as coefficients in (1.26).

Case(b): Boundary value problem (BVP)

For the boundary value problem of (1.21), given y(0), y(1),.....y(m-1) (m initial conditions) and y(N), y(N-1), ... y(N-n+1) (n terminal conditions), from the knowledge of the results of initial value problem, it is apparent that boundary layers occur at k = 0,1,....m-1. Therefore in the BVP also, the degenerate equation which is the same as in IVP loses the m initial conditions at these points. The recovery of these lost initial conditions is therefore accomplished by using the same transformations as in IVP. Consequently, the total series solution (1.26) and the series equations for the outer and correction series are the same as in IVP, that is (1.30) and (1.31) respectively. The selection of the boundary values required to solve these equation is however, different. They are obtained from (1.26) as shown below.

Outer Series:

$h^0 : y_t^{(0)}(N) = y(N), \; y_t^{(0)}(N-1) = y(N-1), \; \ldots$

$\qquad y_t^{(0)}(N-n+1) = y(N-n+1)$

$h^1 : y_t^{(1)}(N) = y_t^{(1)}(N-1) = \ldots = y_t^{(1)}(N-n+1) = 0$

$\ldots \; \ldots \; \ldots \; \ldots \; \ldots \; \ldots \; \ldots \; \ldots \; \ldots \; \ldots \; \ldots$

$h^j : y_t^{(j)}(N) = y_t^{(j)}(N-1) = \ldots = y_t^{(j)}(N-n+1) = 0$

Correction Series:

$\quad$ For $s = m-1, \; m-2, \; \ldots \; 0$

$h^0 : w_s^{(0)}(s) = y(s) - y_t^{(0)}(s)$

$\qquad w_s^{(0)}(0) = w_s^{(0)}(1) \ldots = w_s^{(0)}(s-1) = w_s^{(0)}(s+1) =$

$\qquad\qquad\qquad \ldots = w_s^{(0)}(m-1) = 0$

$h^1 : w_s^{(1)}(s) = - y_t^{(1)}(s)$

$\qquad w_s^{(1)}(0) = w_s^{(1)}(1) = \ldots = w_s^{(1)}(s-1) = w_s^{(1)}(s+1) =$

$\qquad\qquad\qquad \ldots = w_s^{(1)}(m-1) = 0$

$\ldots \; \ldots \; \ldots \; \ldots \; \ldots \; \ldots \; \ldots \; \ldots \; \ldots$

$h^j : w_s^{(j)}(s) = - y_t^{(j)}(s)$

$\qquad w_s^{(j)}(0) = w_s^{(j)}(1) = \ldots = w_s^{(j)}(s-1) = w_s^{(j)}(s+1) \ldots$

$\qquad\qquad\qquad = w_s^{(j)}(m-1) = 0$

Example 1.3

Consider the boundary value problem with the small parameters occurring at the right end as

$$y(k+3) - 1.45\ y(k+2) - 0.09\ y(k+1) + 0.0225\ y(k) = 0 \qquad \ldots \quad (1.32)$$

$$y(0) = 10.0, \ y(1) = 8.0, \ y(8) = 1.0$$

The characteristic roots of (1.32) are

$$m_1 = 0.1, \ m_2 = -0.15, \ m_3 = 1.5$$

Letting h = 0.1, (1.32) is written as

$$y(k+3) - 1.45\ y(k+2) - 0.9\ hy(k+1) + 2.25\ h^2 y(k) = 0$$

Using the method developed above, the outer series and the correction series equations are obtained. The outer series equation is the same as (1.32). Two correction series equations are obtained as below.

$$h\ w_s(k+3) - 1.45 w_s(k+2) - 0.9 w_s(k+1) + 2.25 w_s(k) = 0, \qquad s = 1,2$$
$$\ldots \quad (1.33)$$

where $w_s(k)$ corresponds to the correction for the recovery of the lost initial condition at k = s.

The series equations of (1.32), and (1.33) upto the second order approximation are

Outer Series:

$$h^0 : y^{(0)}(k+3) - 1.45\ y^{(0)}(k+2) = 0 \qquad\qquad \ldots \quad (1.34a)$$

$$h^1 : y^{(1)}(k+3) - 1.45 y^{(1)}(k+2) - 0.9 y^{(0)}(k+1) = 0 \qquad \ldots \quad (1.34b)$$

$$h^2 : y^{(2)}(k+3) - 1.45 y^{(2)}(k+2) - 0.9 y^{(1)}(k+1)$$

$$+ 2.25 y^{(0)}(k) = 0 \qquad \ldots \quad (1.34c)$$

Correction Series:

For s = 0, 1

h^0 : $-1.45w_s^{(0)}(k+2) - 0.9w_s^{(0)}(k+1) + 2.25w_s^{(0)}(k) = 0$ ... (1.35a)

h^1 : $-1.45w_s^{(1)}(k+2) - 0.9w_s^{(1)}(k+1) + 2.25w_s^{(1)}(k)$

$$= - w_s^{(0)}(k+3) \qquad \qquad \qquad ... \quad (1.35b)$$

h^2 : $-1.34w_s^{(2)}(k+2) - 0.9w_s^{(2)}(k+1) + 2.25w_s^{(2)}(k)$

$$= - w_s^{(1)}(k+3) \qquad \qquad \qquad ... \quad (1.35c)$$

The algorithm for solving the series equations (1.34) and (1.35) is given below.

(i) Solve (1.34a) with $y^{(0)}(8) = y(8)$ and find $y^{(0)}(k)$,

   k= 0.1,.....8

(ii) (a) Taking $w_0^{(0)}(0) = y(0) - y^{(0)}(0)$ and $w_0^{(0)}(1) = 0$

   evaluate $w_0^{(0)}(2)$ from (1.35a)

   (b) Taking $w_1^{(0)}(1) = y(1) - y^{(0)}(1)$ and $w_1^{(0)}(0) = 0$,

   find $w_1^{(0)}(2)$ and $w_1^{(0)}(3)$ from (1.35a)

(iii) Solve (1.34b) with $y^{(1)}(8) = 0$, and find $y^{(1)}(k)$,

   k = 0,1,.....8

(iv) (a) Take $w_0^{(1)}(0) = - y^{(1)}(0)$, $w_0^{(1)}(1) = 0$

   (b) Taking $w_1^{(1)}(1) = - y^{(1)}(1)$, $w_1^{(1)}(0)$ 0, find

   $w_1^{(1)}(2)$ from (1.35b)

(v) Solve (1.34c) with $y^{(2)}(8) = 0$, and find $y^{(2)}(k)$,

   k = 0,1,.... 8

   (a) Take $w_0^{(2)}(0) = - y^{(2)}(0)$, and $w_0^{(2)}(1) = 0$

   (b) Take $w_1^{(2)}(1) = - y^{(2)}(1)$, and $w_1^{(2)}(0) = 0$

The total series solution for second-order approximation is

$$y(k) = y^{(0)}(k) + hy^{(1)}(k) + h^2 y^{(2)}(k)$$

$$+ h^k [w_0^{(0)}(k) + hw_0^{(1)}(k) + h^2 w_0^{(2)}(k)]$$

$$+ h^{k-1} [w_1^{(0)}(k) + hw_1^{(1)}(k) + h^2 w_1^{(2)}(k)]$$

The results in Table 1.3 clearly show that the approximate solutions approach the exact solution.

Note: It may be noted from the above algorithm that the correction series is to be solved only for a few value of k. Thus the solution of the correction series is very simple and the various correction terms are found by simple recursion.

Table 1.3

Comparison of the approximate and exact solution of
Example 1.3

k	Degenerate solution	Zeroth order solution	1st order solution	2nd order solution	Exact solution
0	0.0512	10.0000	10.0000	10.0000	10.0000
1	0.0742	8.0000	8.0000	8.0000	8.0000
2	0.1076	0.1076	-0.4120	-0.1430	-0.1599
3	0.1560	0.1560	0.1226	0.2876	0.2632
4	0.2262	0.2262	0.1875	0.2000	0.1872
5	0.3280	0.3280	0.2859	0.2986	0.2988
6	0.4756	0.4756	0.4349	0.4463	0.4442
7	0.6897	0.6897	0.6601	0.6678	0.6667
8	1.0000	1.0000	1.0000	1.0000	1.0000

Note: The degenerate and zeroth order solutions are the same except at k = 0 and k = 1.

Small parameters at the left end: L-type

Consider (1.20) with the n coefficients $c_{m+n}, c_{m+n-1} \cdots \cdots c_{m+1}$ negligibly small in comparison with the others.

Defining $c_{m+n} = h^n$; $c_{m+n-1} = h^{n-1} a_{m+n-1}$; .....

$c_{m+1} = h a_{m+1}$; $c_m = a_m$; $c_{m-1} = a_{m-1}$; .... $c_o = a_o$ and $d_o = b_o$.

Equation (1.20) can be expressed in the general form of L-type as

$$h^n y(k+m+n) + h^{n-1} a_{m+n-1} y(k+m+n-1) + \ldots + h\, a_{m+1} y(k+m+1)$$

$$+ a_m y(k+m) + a_{m-1} y(k+m-1) + \ldots + a_o y(k) = b_o u(k) \qquad \ldots \quad (1.36)$$

The degenerate form of the above equation (1.36) is

$$a_m y^{(0)}(k+m) + a_{m-1} y^{(0)}(k+m-1) + \ldots + a_o y^{(0)}(k) = b_o u(k)$$
$$\qquad \ldots \quad (1.37)$$

As explained in the case of second order equation of L-type, for the general (n+m)th order difference equation (1.36) also, only boundary value problem is of interest. For the boundary value problem given y(0), y(1), ...y(m-1) (m initial conditions) and y(N), y(N-1), ...y(N-n+1) (n terminal conditions) the small parameters appearing at the left end lead to the formation of n number of boundary layers at k = N, N-1, ...N-n+1. As a result, the degenerate equation (1.37) which is of order m satisfies the given m initial conditions and drops the n terminal conditions. In order to recover these lost terminal conditions, we therefore incorporate n separate correction series, each accounting for one lost terminal condition. The transformations for the correction series involve h^{N-k}, h^{N-k-1}, ...,$h^{N-(k+n-1)}$.

The total series solution of (1.36) is written as

$$y(k) = \sum_{j=0}^{\infty} h^j y_t^{(j)}(k) + h^{N-k} \sum_{j=0}^{\infty} h^j w_N^{(j)}(k)$$

$$+ h^{N-k-1} \sum_{j=0}^{\infty} h^j w_{N-1}^{(j)}(k) + \ldots + h^{N-k-(n-1)} \sum_{j=0}^{\infty} h^j w_{N-n+1}^{(j)}(k)$$
$$\qquad \ldots \quad (1.38)$$

where $\sum_{j=0}^{\infty} y_t^{(j)}(k)$ is the <u>outer series</u>, $\qquad \ldots \quad (1.39a)$

and

$$\sum_{j=0}^{\infty} w_s^{(j)}(k), \quad s = N, N-1, \ldots N-n+1, \qquad \ldots (1.39b)$$

is the <u>correction series</u> to recover the terminal conditon $y(s)$.

Substituting the above series solution (1.38) in (1.36) and separating the outer series and the various correction series terms (as in R-type) equations for the outer series and the correction series are obtained.

The equation for the outer series which is identical with the given equation (1.36) is

$$h^n y_t(k+m+n) + h^{n-1} a_{m+n-1} y_t(k+m+n-1) + \ldots + h\, a_{m+1} y_t(k+m+1)$$

$$+ a_m y(k+m) + \ldots + a_o y(k) = b_o u(k)$$

The equation of the correction series is

$$w_s(k+m+n) + a_{m+n-1} w_s(k+m+n-1) + \ldots + a_{m+1} w_s(k+m+1)$$

$$+ a_m w_s(k+m) + h\, a_{m-1} w_s(k+m-1) + h^2 a_{m-2} w_s(k+m-2)$$

$$+ \ldots + h^m a_o w_s(k) = 0,$$

$$s = N, N-1, \ldots N-n+1.$$

Assuming series expansions as in (1.39), the equations for zeroth, first and higher order approximations can be obtained as shown below.

Outer Series

$$h^0 : a_m y_t^{(0)}(k+m) + a_{m-1} y_t^{(0)}(k+m-1) + \ldots + a_o y_t^{(0)}(k) = b_o\, u(k)$$

$$h^1 : a_m y_t^{(1)}(k+m) + a_{m-1} y_t^{(1)}(k+m-1) + \ldots$$

$$+ a_o y_t^{(1)}(k) = - a_{m+1}\, y_t^{(0)}(k+m+1)$$

$$h^j : a_m y_t^{(j)}(k+m) + a_{m-1}\, y_t^{(j)}(k+m-1) + \ldots + a_o y_t^{(j)}(k) =$$

$$- a_{m+1} y_t^{(j-1)}(k+m+1) - a_{m+2} y_t^{(j-2)}(k+m+2)$$

$$\ldots - a_{m+j}\, y_t^{(0)}(k+m+j) \qquad\qquad \ldots \quad (1.40)$$

Correction Series

For $s = N,\ N-1,\ \ldots,\ N - n + 1$

$$h^0 : w_s^{(0)}(k+m+n) + a_{m+n-1}\, w_s^{(0)}(k+m+n-1) + \ldots$$

$$+ a_m\, w_s^{(0)}(k+m) = 0$$

$$h^1 : w_s^{(1)}(k+m+n) + a_{m+n-1}\, w_s^{(1)}(k+m+n-1) + \ldots$$

$$+ a_m\, w_s^{(1)}(k+m) = - a_{m-1}\, w_s^{(0)}(k+m-1)$$

$$\ldots \quad \ldots \quad \ldots \quad \ldots \quad \ldots \quad \ldots$$

$$h^j : w_s^{(j)}(k+m+n) + a_{m+n-1}\, w_s^{(j)}(k+m+n-1) + \ldots$$

$$+ a_m w_s^{(j)}(k+m) = - a_{m-1} w_s^{(j-1)}(k+m-1)$$

$$- a_{m-2}\, w_s^{(j-2)}(k+m-2) - \ldots - a_{m-j}\, w_s^{(0)}(k+m-j)$$

$$\ldots \quad (1.41)$$

The boundary conditions required to solve the above series equations are readily obtained from (1.38) as shown

Outer Series:

$$h^0 : y_t^{(0)}(0) = y(0) , \; y_t^{(0)}(1) = y(1), \; \ldots, \; y_t^{(0)}(m-1) = y(m-1)$$

$$h^1 : y_t^{(1)}(0) = y_t^{(1)}(1) = \ldots = y_t^{(1)}(m-1) = 0$$

$$\cdots \qquad \cdots \cdots \cdots \cdots \cdots \cdots \cdots$$

$$h^j : y_t^{(j)}(0) = y_t^{(j)}(1) = \ldots = y_t^{(j)}(m-1) = 0$$

Correction Series:

For $s = N-n+1, \; N-n+2, \; \ldots, \; N$

$$h^0 : w_s^{(0)}(s) = y(s) - y_t^{(0)}(s)$$

$$w_s^{(0)}(N-n+1) = w_s^{(0)}(N-n+2) = \ldots = w_s^{(0)}(s-1)$$

$$= w_s^{(0)}(s+1) = \ldots = w_s^{(0)}(N) = 0$$

$$h^1 : w_s^{(1)}(s) = - y_t^{(1)}(s)$$

$$w_s^{(1)}(N-n+1) = w_s^{(1)}(N-n+2) = \ldots = w_s^{(1)}(s-1)$$

$$= w_s^{(1)}(s+1) = \ldots = w_s^{(1)}(N) = 0$$

$$\cdots \cdots \cdots \cdots \cdots \cdots \cdots \cdots \cdots$$

$$h^j : w_s^{(j)}(s) = - y_t^{(j)}(s)$$

$$w_s^{(j)}(N-n+1) = w_s^{(j)}(N-n+2) = \ldots = w_s^{(j)}(s-1)$$

$$= w_s^{(j)}(s+1) = \ldots = w_s^{(j)}(N) = 0$$

Example 1.4

Consider a third order boundary value problem with the small parameters at the left end (L-type) as

$$y(k+3) - 18.3y(k+2 + 89y(k+1) - 60y(k) = 10 \ u(k)$$

... (1.42)

$$y(10) = 10.0; \ y(9) = 8.0; \ y(0) = 2.5, \ u(k) = 1$$

The characteristic roots of (1.42) are

$$m_1 = 10.0; \ m_2 = 7.5; \ m_3 = 0.8$$

Letting h = 0.1, (1.42) is rewritten in the standard form as

$$h^2 y(k+3) - 1.83hy(k+2) + 0.89y(k+1) - 0.6y(k) = 0.1 \ u(k) \ \text{... (1.43)}$$

Using the method developed above, the outer series and correction series equations are obtained. The outer series equation is same as (1.42). Two correction series equations are obtained as below.

For s = 9, 10

$$w_s(k+3) - 1.83w_s(k+2) + 0.89w_s(k+1) - 0.6hw_s(k) = 0 \qquad \text{... (1.44)}$$

where $w_s(k)$ corresponds to the recovery of the lost boundary condition at k = s. The series equations of (1.42) and (1.44) upto second order approximation are

Outer Series

$$h^0 : 0.89y^{(0)}(k+1) - 0.6y^{(0)}(k) = 0.1 \ u(k) \qquad \text{... (1.45a)}$$

$$h^1 : 0.89y^{(1)}(k+1) - 0.6y^{(1)}(k) = 1.83y^{(0)}(k+2) \qquad \text{... (1.45b)}$$

$$h^2 : 0.89y^{(2)}(k+1) - 0.6y^{(2)}(k) = 1.83y^{(1)}(k+2) - y^{(0)}(k+3)$$

... (1.45c)

Correction Series

For $s = 9,\ 10$

h^0 : $w_s^{(0)}(k+3) - 1.83w_s^{(0)}(k+2) + 0.89w_s^{(0)}(k+1) = 0$ ... (1.46a)

h^1 : $w_s^{(1)}(k+3) - 1.83w_s^{(1)}(k+2) + 0.89w_s^{(0)}(k+1) = 0.6\ w_s^{(0)}(k)$

          ... (1.46b)

h^2 : $w_s^{(2)}(k+3) - 1.83w_s^{(1)}(k+2) + 0.89\ w_s^{(2)}(k+1) = 0.6\ w_s^{(1)}(k)$

          ... (1.46c)

The series equations given above are solved step by step as shown below:

(i) Solve (1.45a) with $y^{(0)}(0) = y(0)$, and find $y^{(0)}(k)$,

     $k = 0,1,\ ......\ 10$

(ii) (a) Taking $w_9^{(0)}(9) = y(9) - y^{(0)}(9)$, and $w_9^{(0)} = 0$,

     evaluate $w_9^{(0)}(8)$ and $w_9^{(0)}(7)$ from (1.46a)

     (b) Taking $w_{10}^{(0)}(10) = -y^{(0)}(10)$ and $w_{10}^{(0)}(9) = 0$,

     evaluate $w_{10}^{(0)}(8)$ from (1.46a)

(iii) Solve (1.45b) with $y^{(1)}(0) = 0$, and find $y^{(1)}(k)$,

     $k = 0,\ 1,\ .....,\ 10$

(iv) (a) Taking $w_9^{(1)}(9) = -\ y^{(1)}(9)$, $w_9^{(1)}(10) = 0$,

     evaluate $w_9^{(1)}(8)$ from (1.46b)

     (b) Take $w_{10}^{(1)}(10) = -\ y^{(1)}(10)$, $w_{10}^{(1)}(9) = 0$

(v) Solve (1.45c) with $y^{(2)}(0) = 0$, and find $y^{(2)}(k)$,

     $k = 0,\ 1,\ .....,\ 10$

(vi) (a) Take $w_9^{(2)}(9) = -\ y_9^{(2)}(9)$, $w_9^{(2)}(10) = 0$

     (b) Take $w_{10}^{(2)}(10) = -\ y_{10}^{(2)}(10)$, $w_{10}^{(2)}(9) = 0$

The total series solution is

$$y(k) = y^{(0)}(k) + h\, y^{(1)}(k) + h^2\, y^{(2)}(k) + \ldots$$

$$+ n^{N-k-1}\, [w_9^{(0)}(k) + h\, w_9^{(1)}(k) + h^2 w_9^{(2)} + \ldots]$$

$$+ h^{N-k}[w_{10}^{(0)}(k) + h\, w_{10}^{(1)}(k) + h^2\, w_{10}^{(2)}(k) + \ldots]$$

The results in Table 1.4 clearly indicate that the series solutions approach the exact solution very closely as the order of approximation is increased.

1.4 Two-Time-Scale Property

An examination of the characteristic roots of the R-and L-type equations reveals the two-time-scale property of the singularly perturbed difference equations. For the second order R-type equation, (1.1) the two characteristic roots arc given by (1.2). For sufficiently small values of h, they are written as

$$m_1 = - a + \frac{h}{a} + 0\ (h^2) \qquad\qquad \ldots \ (1.47a)$$

$$m_2 = - \frac{h}{a} + 0\ (h^2) \qquad\qquad \ldots \ (1.47b)$$

TABLE 1.4

Comparison of approximate and exact solutions of Example 1.4

k	Degenerate solution	Zeroth order solution	1st order solution	2nd order solution	Exact solution
0	2.5000	2.5000	2.5000	2.5000	2.5000
1	1.7979	1.7978	2.0701	2.1390	2.1709
2	1.3243	1.3243	1.7146	1.8397	1.9077
3	1.0052	1.0052	1.4307	1.5940	1.6972
4	0.7900	0.7900	1.2095	1.3947	1.5295
5	0.6449	0.6449	1.0402	1.2349	1.3998
6	0.5472	0.5472	0.9126	1.1080	1.3260
7	0.4812	0.4812	0.8174	1.2441	1.4565
8	0.4368	0.4368	2.3084	2.4837	2.6269
9	0.4068	8.0000	8.0000	8.0000	8.0000
10	0.3866	10.0000	10.0000	10.0000	10.0000

Note: The degenerate and zeroth order solutions are the same except at k = 9 and k = 10

The solution of (1.1) is written from (1.3) as

$$y(k) = y_1(k) + y_2(k)$$

where $y_1(k) = c_1(m_1)^k$ and $y_2(k) = c_2(m_2)^k$

From (1.47), it follows that $|m_2| \ll |m_1|$. As a result, $y_2(k)$ has a much faster variation than $y_1(k)$ with respect to k. Therefore $y_1(k)$ and $y_2(k)$ are called the "slow" and "fast" modes of $y(k)$ and (1.1) is said to possess the two-time-scale property. As h tends to zero in the degeneration process, it is clear from (1.47) that the root m_1 corresponding to the slow mode approaches - a, while the root m_2 corresponding to the fast mode approaches zero.

The characteristic roots of the L-type difference equation (1.16) are given by (1.17). For sufficiently small values of h, they can be written as

$$m_1 = -\frac{1}{a} + 0(h) \qquad \qquad \ldots \text{(1.48a)}$$

$$m_2 = -\frac{a}{h} + \frac{1}{a} + 0(h) \qquad \qquad \ldots \text{(1.48b)}$$

The solution of (1.16) can be written from (1.18) as

$$y(k) = y_1(k) + y_2(k)$$

where $y_1(k) = c_1(m_1)^k$ and $y_2(k) = c_2(m_2)^k$

From (1.48), $|m_2| \gg |m_1|$ and it follows that $y_2(k)$ has a much faster variation than $y_1(k)$ with respect to k. Therefore for this type of difference equation $y_1(k)$ and $y_2(k)$ are called as the "slow" and "fast" modes respectively. From (1.48), it is clear that in the degeneration process, the root m_1 corresponding to the slow mode approaches $-\frac{1}{a}$ while the root m_2 corresponding to the fast mode approaches infinity.

Thus the main difference in the two-time-scale properties of the R- and L-types is in respect of the behaviour of their fast modes. In the R-type, the fast mode is always stable and corresponds to a fast

<u>decay</u> of the solution near k = 0. But in L-type, the fast mode is always <u>unstable</u> and corresponds to a <u>fast rise</u> near k = N (terminal point). However, in both the types the roots corresponding to the slow mode are in the neighborhood of unity. Thus their slow modes can be stable or unstable.

We can extend the above ideas to the higher order difference equations. The R-type (n+m)th order difference equation (1.21) evidently has the "<u>slow</u>" modes with n characteristic roots that lie in the neighborhood of unity. It has "<u>fast</u>" modes with m characteristic roots whose moduli are much smaller than unity and <u>tend to zero</u> in the degeneration process. Similarly, an (n+m)th order L-type difference equation (1.36) has slow modes with m roots in the neighborhood of unity and fast modes with n roots whose moduli are much larger than unity and <u>approach infinity</u> in the degeneration process.

1.5 Conclusions

As a first step towards the singular perturbation analysis of discrete control systems, in this Chapter the difference equations are considered in the classical form. The special features of this chapter are:
   (i) Extension of the results of Comstock and Hsiao [18] to obtain series expansions for any order of approximation, and
   (ii) The analysis of higher order difference equations.

It is worthwhile to examine some aspects of singular perturbations in difference equations along with those in differential equations.
   (i) In difference equations, the small parameters appearing either at the left end or at the right end gives the singularly perturbed structure, whereas in differential equations the singular perturbation structure is obtained only when the small parameter is at the left end, that is, multiplying the highest derivatives.
   (ii) The location of the boundary layer in difference equations depends on the position (left or right end) of the singular perturbation parameter. But in differential equations, the signs of the coefficients decide the location of the boundary layer [8].
   (iii) In higher order difference equations with several small parameters, number of boundary layers occur at several points, that is, at k = 0, 1, ... or k = N, N-1,..... whereas in differential equations the boundary layer occurs at the initial, final, or at both the points.

(iv) For difference equations both the outer and correction series equations are of reduced order, while in differential equations only the outer series equations are of reduced order.

(v) A very important feature in difference equations is that the evaluation of boundary conditions required to solve the various series equations is simple and straight-forward. In differential equations this is often a formidable task [8].

(vi) In difference equations as the correction series is multiplied with coefficients like h^k and h^{N-k}, it is enough if the correction series equations are solved for a few values of k near the boundary layer. This aspect coupled with the reduction of the order of the correction series equations offers considerable simplicity.

CHAPTER - 2

MODELLING AND ANALYSIS OF SINGULARLY PERTURBED DIFFERENCE
EQUATIONS IN STATE SPACE FORM

In this Chapter, investigations are carried out regarding the
modelling and analysis of singularly perturbed difference equation in
state space form which is in a general framework suitable for control
engineers.   Three state space discrete models are formulated.   The
various implications involved in obtaining discrete models for continuous and sampled-data systems are examined.   Singular perturbation
methods are developed where the approximate solutions are sought in
terms of an outer series obtained from the low-order model of the
original system and correction series obtained from the low-order model
of the transformed system.   The asymptotic correctness of the series
expansions is established.   Examples are provided to illustrate the
proposed methods [29-32].

2.1 State Space Formulation

(a) Small parameter at the right end

Consider the general (n+m)th order linear, shift-invariant, difference equation with small parameters occuring at the right end as

$$y(k+m+n) + a_{m+n-1}y(k+m+n-1) + \ldots + a_m y(k+m)$$

$$+ a_{m-1}hy(k+m-1) + \ldots + a_1 h^{m-1}y(k+1) + a_0 h^m y(k) = b_0 u(k)$$

$$\ldots (2.1)$$

with h as the small parameter.

Defining the state variables as

$x_1(k) = y(k+m)$

$x_2(k) = y(k+m+1)$

. . .   .

$x_n(k) = y(k+m+n-1)$, and

$z_1(k) = h^{m-1}y(k)$

$z_2(k) = h^{m-2}y(k+1)$

. . .   .

$z_m(k) = y(k+m-1)$

The following first order difference equations are written as

$x_1(k+1) = x_2(k)$

$x_2(k+1) = x_3(k)$

. . .   . . .

$x_n(k+1) = - a_{m+n-1}x_n(k) - a_{m+n-2}x_{n-1}(k) - \ldots - a_m x_1(k)$

$$- a_{m-1}hz_m(k) - \ldots - a_1 hz_2(k) - a_0 hz_1(k)$$

$$+ b_o u(k)$$

$z_1(k+1) = hz_2(k)$

$z_2(k+1) = hz_3(k)$

. . .   . . .

$z_m(k+1) = x_1(k)$

The above first order equations are rearranged and written in matrix form as in (2.2).

$$
\begin{bmatrix} x_1(k+1) \\ x_2(k+1) \\ \cdot \\ x_n(k+1) \\ z_1(k+1) \\ \cdot \\ z_m(k+1) \end{bmatrix}
=
\begin{bmatrix}
0 & 1 & 0 & \cdots & 0 & 0 & 0 & \cdots & 0 \\
0 & 0 & 1 & \cdots & 0 & 0 & 0 & \cdots & 0 \\
\cdot & \cdot & \cdot & \cdots & \cdot & \cdot & \cdot & \cdots & \cdot \\
-a_m & -a_{m+1} & -a_{m+2} & \cdots & -a_{m+n-1} & -a_0 h & -a_1 h & \cdots & -a_{m-1}h \\
0 & 0 & 0 & \cdots & 0 & 0 & 0 & \cdots & 0 \\
\cdot & \cdot & \cdot & \cdots & \cdot & \cdot & \cdot & \cdots & \cdot \\
1 & 0 & 0 & \cdots & 0 & 0 & 0 & \cdots & 0
\end{bmatrix}
\begin{bmatrix} x_1(k) \\ x_2(k) \\ \cdot \\ x_n(k) \\ z_1(k) \\ \cdot \\ z_m(k) \end{bmatrix}
+
\begin{bmatrix} 0 \\ 0 \\ \cdot \\ b_0 \\ 0 \\ \cdot \\ 0 \end{bmatrix}
u(k)
\qquad \cdots \ (2.2)
$$

This matrix difference equation (2.2) is partitioned as

$$
\begin{bmatrix} x(k+1) \\ z(k+1) \end{bmatrix}
=
\begin{bmatrix} A_{11} & hA_{12} \\ A_{21} & hA_{22} \end{bmatrix}
\begin{bmatrix} x(k) \\ z(k) \end{bmatrix}
+
\begin{bmatrix} E \\ F \end{bmatrix}
u(k)
\qquad \cdots \ (2.3)
$$

where x(k) is (nx1) state vector;

z(k) is (mx1) state vector;

$$A_{11} = \begin{bmatrix} 0 & 0 & 0 & \cdots & 0 & 0 \\ 0 & 0 & 0 & \cdots & 0 & 0 \\ \vdots & \vdots & \vdots & \cdots & \vdots & \cdot \\ -a_m & -a_{m+1} & -a_{m+2} & \cdots & -a_{m+n-1} \end{bmatrix}$$
(nxn)

$$A_{12} = \begin{bmatrix} 0 & 0 & 0 & \cdots & 0 & 0 \\ 0 & 0 & 0 & \cdots & 0 & 0 \\ \vdots & \vdots & \vdots & \cdots & \vdots & \cdot \\ -a_0 & -a_1 & \cdots & -a_{m-2} & -a_{m-1} \end{bmatrix}$$
(nxm)

$$A_{21} = \begin{bmatrix} 0 & 0 & \cdots & 0 & 0 \\ 0 & 0 & \cdots & 0 & 0 \\ \vdots & \vdots & \cdots & \vdots & \vdots \\ 1 & 0 & \cdots & 0 & 0 \end{bmatrix}$$
(mxn)

$$A_{22} = \begin{bmatrix} 0 & 0 & 1 & \cdots & 0 & 0 \\ 0 & 0 & 0 & \cdots & 1 & 0 \\ \vdots & \vdots & \vdots & \cdots & \vdots & \vdots \\ 0 & 0 & 0 & \cdots & 0 & 0 \end{bmatrix}$$
(mxm)

$$E = \begin{bmatrix} b_0 & \cdots & \cdot & 0 & 0 \end{bmatrix}$$
(nx1)

$$E = \begin{bmatrix} 0 & \cdots & \cdot & 0 & 0 \end{bmatrix}$$
(mx1)

The state space model (2.3) is called as the C-model, since the small parameter appears in the column of the system matrix.

Defining the state variables in an alternate way for (2.1) as

$$x_1(k) = y(k+m)$$

$$x_2(k) = y(k+m+1)$$

...   .

$$x_n(k) = y(k+m+n-1)$$

$$z_1(k) = h^m y(k)$$

$$z_2(k) = h^{m-1} y(k+1)$$

...   .

$$z_m(k) = h \, y(k+m-1)$$

the following first order difference equations are written from (2.1)

$$x_1(k+1) = x_2(k)$$

$$x_2(k+1) = x_3(k)$$

...   .

$$x_n(k+1) = - a_{m+n-1} \, x_n(k) - a_{m+n-2} \, x_{n-1}(k) - \dots - a_m x_1(k)$$

$$- a_{m-1} z_m(k) - a_{m-2} z_{m-1}(k) - \dots - a_0 z_1(k) + b_0 u(k)$$

$$z_1(k+1) = h \, z_2(k)$$

$$z_2(k+1) = h \, z_3(k)$$

...   ...

$$z_m(k+1) = h \, x_1(k)$$

The above equations are expressed in matrix form as

$$
\begin{bmatrix} x_1(k+1) \\ x_2(k+1) \\ \vdots \\ x_n(k+1) \\ z_1(k+1) \\ z_2(k+1) \\ \vdots \\ z_m(k+1) \end{bmatrix}
=
\begin{bmatrix}
0 & 1 & 0 & \cdots & & & & & \\
0 & 0 & 1 & \cdots & & & & & \\
\vdots & & & & & & & & \\
-a_m & -a_{m+1} & -a_{m+2} & \cdots & -a_{m+n-1} & -a_o & -a_1 & \cdots & -a_{m-1} \\
0 & 0 & 0 & & 0 & 0 & h & & 0 \\
0 & 0 & 0 & & 0 & 0 & 0 & h & 0 \\
\vdots & & & & & & & & \\
1 & 0 & 0 & & 0 & 0 & 0 & \cdots & 0
\end{bmatrix}
\begin{bmatrix} x_1(k) \\ x_2(k) \\ \vdots \\ x_n(k) \\ z_1(k) \\ z_2(k) \\ \vdots \\ z_m(k) \end{bmatrix}
+
\begin{bmatrix} 0 \\ 0 \\ \vdots \\ b_o \\ 0 \\ 0 \\ \vdots \\ 0 \end{bmatrix} u(k)
\qquad (2.4a)
$$

This matrix difference equation (2.2) is partitioned as

$$
\begin{bmatrix} x(k+1) \\ z(k+1) \end{bmatrix}
=
\begin{bmatrix} A_{11} & A_{12} \\ hA_{21} & hA_{22} \end{bmatrix}
\begin{bmatrix} x(k) \\ z(k) \end{bmatrix}
+
\begin{bmatrix} E \\ hF \end{bmatrix} u(k)
\qquad (2.4b)
$$

where $x(k)$, $z(k)$, A_{11}, A_{12}, A_{21}, A_{22}, E and F are the same as given for the C-model (2.3)..

The state space model (2.4b) is called the R-model, as the small parameter h appears in the row of the system matrix.

Note: In the above model (2.4b) if z(k) is replaced with hz(k) the resulting equation is the same as the C-model (2.3). Similarly in C-model (2.3), if z(k) is replaced with z(k)/h, R-model (2.4) is obtained.

(b) Small parameters at the left end

Consider the general (n+m)th order difference equation with the small parameters at the left end

$$h^n y(k+m+n)+h^{n-1}a_{m+n-1} \, y(k+m+n-1) +\ldots+h \, a_{m+1} \, y(k+m+1)$$

$$+a_m y(k+m)+\ldots+a_0 y(k) = b_0 u(k) \qquad \ldots \quad (2.5)$$

Defining the state variables as

$$x_1(k) \;=\; y(k)$$

$$x_2(k) \;=\; y(k+1)$$

$$\ldots \qquad \ldots$$

$$x_m(k) \;=\; y(k+m-1)$$

$$z_1(k) \;=\; y(k+m)$$

$$z_2(k) \;=\; hy(k+m+1)$$

$$\ldots \qquad \ldots$$

$$z_n(k) \;=\; h^{n-1} \, y(k+m+n-1)$$

the following first order difference equations are written from (2.5)

$$x_1(k+1) \;=\; x_2(k)$$

$$x_2(k+1) \;=\; x_3(k)$$

$$\cdots \qquad \cdots$$

$$x_m(k+1) \;=\; z_1(k)$$

$$hz_1(k+1) \;=\; z_2(k)$$

$$hz_2(k+1) \;=\; z_3(k)$$

$$\cdots \qquad \cdots$$

$$hz_n(k+1) \;=\; -a_{m+n-1}\, z_n(k) - a_{m+n-2} z_{n-1}(k) - \cdots - a_{m+1} z_2(k)$$

$$-a_m z_1(k) - a_{m-1}\, x_m(k) - \cdots - a_0 x_1(k) + b_0 u(k)$$

The above equations can be arranged in the matrix form as in (2.6a)

$$
\begin{bmatrix}
x_1(k+1) \\
x_2(k+1) \\
\cdots \\
x_m(k+1) \\
hz_1(k+2) \\
hz_2(k+1) \\
\cdots \\
hz_n(k+1)
\end{bmatrix}
=
\begin{bmatrix}
0 & 1 & 0 & \cdots & 0 & 0 & 0 & \cdots & 0 \\
0 & 0 & 1 & \cdots & 0 & 0 & 0 & \cdots & 0 \\
\cdot & \cdot & \cdot & \cdots & \cdot & \cdot & \cdot & \cdots & \cdot \\
0 & 0 & 0 & \cdots & 1 & 0 & 0 & \cdots & 0 \\
0 & 0 & 0 & \cdots & 0 & 0 & 1 & \cdots & 0 \\
0 & 0 & 0 & \cdots & 0 & 0 & 0 & \cdots & 0 \\
\cdot & \cdot & \cdot & \cdots & \cdot & \cdot & \cdot & & \cdot \\
-a-a_0-1_1-a_2 & \cdots & & -a_{m-1}-a_{m+1} & \cdots & & -a_{m+n-1}
\end{bmatrix}
\begin{bmatrix}
x_z(k) \\
x_2(k) \\
\cdots \\
x_m(k) \\
z_1(k) \\
z_2(k) \\
\cdots \\
z_n(k)
\end{bmatrix}
+
\begin{bmatrix}
0 \\
0 \\
\cdot \\
0 \\
0 \\
0 \\
\cdot \\
b_0
\end{bmatrix}
u(k)
$$

$$\cdots \quad (2.6a)$$

and partitioned as

$$
\begin{bmatrix}
x(k+1) \\
hz(k+1)
\end{bmatrix}
=
\begin{bmatrix}
A_{11} & A_{12} \\
A_{21} & A_{22}
\end{bmatrix}
\begin{bmatrix}
x(k) \\
z(k)
\end{bmatrix}
+
\begin{bmatrix}
E \\
F
\end{bmatrix}
u(k)
\qquad \cdots \quad (2.6b)
$$

where $x(k)$ is $(m \times 1)$ state vector;

$z(k)$ is $(n \times 1)$ state vector;

$$A_{11} \atop (m \times m) = \begin{bmatrix} 0 & 1 & 0 & \cdots & 0 \\ 0 & 0 & 1 & \cdots & 0 \\ \cdot & \cdot & \cdot & \cdots & \cdot \\ 0 & 0 & 0 & \cdots & 1 \end{bmatrix} ; \quad A_{12} \atop (m \times n) = \begin{bmatrix} 0 & 0 & \cdots & 0 \\ 0 & 0 & \cdots & 0 \\ \cdot & \cdot & \cdots & \cdot \\ 0 & 0 & \cdots & 0 \end{bmatrix}$$

$$A_{21} \atop (n \times m) = \begin{bmatrix} 0 & 0 & 0 & \cdots & 0 \\ 0 & 0 & 0 & \cdots & 0 \\ \cdot & \cdot & \cdot & \cdots & 0 \\ -a_0 & -a_1 & -a_2 & \cdots & -a_{m-1} \end{bmatrix} ; \quad A_{22} \atop (n \times n) = \begin{bmatrix} 0 & 1 & 0 & \cdots & 0 \\ 0 & 0 & 1 & \cdots & 0 \\ \cdot\cdot & \cdot\cdot & \cdot\cdot & \cdot\cdot & \cdot \\ -a_m & -a_{m+1} & -a_{m+2} & \cdots & -a_{m+n-1} \end{bmatrix}$$

The state space model (2.6b) is called the D-model on account of its similarity with the model of the singularly perturbed differential equation.

2.2 Test for Two-Time-Scale Property

Consider the linear, shift-invariant discrete system described by the difference equation

$$y(k+1) = A\ y(k) + B\ u(k)$$

where $y(k)$ is $(n+m) \times 1$ state vector,

$u(k)$ is $(r \times 1)$ control vector and

A and B are real matrices of compatible dimensionality.

Generally one may not be able to partition the system matrix A into any of the three models described previously, though the system possesses the two-time-scale property. In fact, it is not even apparent whether the system has this property or not. One sure way of assertaining this property is of course to determine the eigenvalues of the system matrix and verify that there are two clusters of eigenvalues widely separated from each other. But this is very difficult and cumbersome particularly for higher order systems. R. G. Phillips has given a method to test the two-time-scale property and bring the system model into one of the two C- or R-models [25]. Following this method, once the two-time-scale property is tested and the system equations are obtained in the partitioned form as

$$\begin{bmatrix} x(k+1) \\ z(k+1) \end{bmatrix} = \begin{bmatrix} A_{11} & A_{12} \\ A_{21} & A_{22} \end{bmatrix} \begin{bmatrix} x(k) \\ z(k) \end{bmatrix} + \begin{bmatrix} B_1 \\ B_2 \end{bmatrix} u(k) \quad \ldots \text{(2.7)}$$

it is easy to identify whether the above system belongs to the C- or R-model. This is done on the basis of the norms of A_{11}, A_{12}, A_{21} and A_{22} where the norm of a matrix is defined as

$$\| A \| = p_{max} (A'A)^{0.5}$$

In (2.7), if $\| A_{22} \|$ and $\| A_{12} \|$ are very small in comparison with $\| A_{11} \|$ and $\| A_{21} \|$, the system is modelled into C-model as shown below.

$$\begin{bmatrix} x(k+1) \\ z(k+1) \end{bmatrix} = \begin{bmatrix} A_{11} & h\bar{A}_{12} \\ A_{21} & h\bar{A}_{22} \end{bmatrix} \begin{bmatrix} x(k) \\ z(k) \end{bmatrix} + \begin{bmatrix} B_1 \\ B_2 \end{bmatrix} u(k)$$

where $h\bar{A}_{12} = A_{12}$ and $h\bar{A}_{22} = A_{22}$

If $\| A_{22} \|$ and $\| A_{21} \|$ are very small in comparison with $\| A_{11} \|$ and $\| A_{12} \|$, the system is modelled into R-model as shown below.

$$\begin{bmatrix} x(k+1) \\ z(k+1) \end{bmatrix} = \begin{bmatrix} A_{11} & A_{12} \\ h\bar{A}_{21} & h\bar{A}_{22} \end{bmatrix} \begin{bmatrix} x(k) \\ z(k) \end{bmatrix} + \begin{bmatrix} B_1 \\ h\bar{B}_2 \end{bmatrix} u(t)$$

where $h\bar{A}_{21} = A_{21}$; $h\bar{A}_{22} = A_{22}$ and $h\bar{B}_2 = B_2$

2.3 Discrete Models of Singularly Perturbed Continuous Systems

Consider the singularly perturbed continuous system described by the differential equation

$$\begin{bmatrix} \dot{x}(t) \\ \varepsilon \dot{z}(t) \end{bmatrix} = \begin{bmatrix} G_{11} & G_{12} \\ G_{21} & G_{22} \end{bmatrix} \begin{bmatrix} x(t) \\ z(t) \end{bmatrix} + \begin{bmatrix} H_1 \\ H_2 \end{bmatrix} u(k)$$

or

$$
\begin{bmatrix} x(t) \\ \\ z(t) \end{bmatrix} = \begin{bmatrix} G_{11} & G_{12} \\ \\ \dfrac{G_{21}}{\varepsilon} & \dfrac{G_{22}}{\varepsilon} \end{bmatrix} \begin{bmatrix} x(t) \\ \\ z(t) \end{bmatrix} + \begin{bmatrix} H_1 \\ \\ \dfrac{H_2}{\varepsilon} \end{bmatrix} u(t) \quad \cdots \quad (2.8)
$$

where x(t) is (nx1) state vector;

z(t) is (mx1) state vector;

u(t) is (rx1) control vecto;

G_{ij}, H_i, i = 1,2, j = 1,2 are of compatible dimensionality.

The discrete model of (2.8) is [42]

$$
\begin{bmatrix} x(k+1) \\ \\ z(k+1) \end{bmatrix} = \begin{bmatrix} A_{11} & A_{12} \\ \\ A_{21} & A_{22} \end{bmatrix} \begin{bmatrix} x(k) \\ \\ z(k) \end{bmatrix} + \begin{bmatrix} E \\ \\ F \end{bmatrix} u(k) \quad \cdots \quad (2.9)
$$

where $\begin{bmatrix} A_{11} & A_{12} \\ A_{21} & A_{22} \end{bmatrix} = e^{GT}$ and $\begin{bmatrix} E \\ F \end{bmatrix} = \int_0^T e^{Gt} \, dt \; H$

with $G = \begin{bmatrix} G_{11} & G_{12} \\ \\ \dfrac{G_{21}}{\varepsilon} & \dfrac{G_{22}}{\varepsilon} \end{bmatrix}$; $H = \begin{bmatrix} H_1 \\ \\ \dfrac{H_2}{\varepsilon} \end{bmatrix}$;

and T is the discretizing interval. The input u(t) is assumed to be constant between any two consecutive discretizing intervals.

Whether or not the discrete model (2.9) is in the singularly perturbed form depends on the relative norms of A_{11} , A_{12} , A_{21} , and A_{22} which in turn depend upon the discretizing interval T. In other words, though the differential equation (2.8) is in the singularly perturbed form, its discrete model is not automatically obtained in this form unless T is appropriately chosen.

Let us take the separation of the two sets of eigenvalues associated with the fast and slow modes as the criterion for the two-time-
scale property and the consequent singular perturbation structure. We
know that in the continuous system, if m_f and m_s are the two typical
eigenvalues (modulus of m_f is the smallest and modulus of m_s is the
largest among the eigenvalues of the fast and the slow modes
respectively) such that the modulus of m_f is much larger than m_s, the
system is said to have the two-time-scale property and can therefore be
cast in the singularly perturbed form (2.8). Using the same criterion,
an approximate formula for the selection of the discretizing interval T
is obtained so that the corresponding discrete model is also in the
singularly perturbed form.

For the sake of simplicity, let us take a second order singularly
perturbed differential equation.

$$
\begin{bmatrix} x(t) \\ \\ z(t) \end{bmatrix}
=
\begin{bmatrix} g_{11} & g_{12} \\ \\ \dfrac{g_{21}}{\varepsilon} & \dfrac{g_{22}}{\varepsilon} \end{bmatrix}
\begin{bmatrix} x(t) \\ \\ z(t) \end{bmatrix}
\qquad \ldots \quad (2.10)
$$

Case (a): Stable fast mode

Let us consider the case when g_{22} is negative in (2.10), that is,
when the fast mode is stable. The stability of the slow mode depends
on the other elements of the system matrix. Let the eigenvalues corresponding to the fast and slow modes be $- m_f$ and $\pm m_s$ ($+ m_s$ and
$- m_s$ correspond to the unstable and stable eigenvalues respectively).
From the theory of singular perturbation analysis of continuous systems
[8], we know that

$$
\left| \frac{m_s}{m_f} \right| \tilde{} \varepsilon, \text{ where } \varepsilon \text{ is the small parameter in (2.10).}
$$

The discrete model of (2.10) is obtained as in (2.9), that is,

$$
\begin{bmatrix} x(k+1) \\ \\ z(k+1) \end{bmatrix}
=
\begin{bmatrix} a_{11} & a_{12} \\ \\ a_{21} & a_{22} \end{bmatrix}
\begin{bmatrix} x(k) \\ \\ z(k) \end{bmatrix}
\qquad \ldots \quad (2.11)
$$

where

$$\begin{bmatrix} a_{11} & a_{12} \\ a_{21} & a_{22} \end{bmatrix} = e^{GT}, \text{ with } G = \begin{bmatrix} g_{11} & g_{12} \\ \dfrac{g_{21}}{\varepsilon} & \dfrac{g_{22}}{\varepsilon} \end{bmatrix},$$

and T as the discretizing interval.

Using Frobenius theorem [42], the eigenvaues of the discrete model are

$$P_f = e^{-m_f T} \;;\qquad P_s = e^{\pm m_s T}$$

where P_f and P_s are the eigenvalues of (2.11) corresponding to the fast and slow modes of the discrete version respectively.

If T is selected such that $e^{-m_f T}$ is much smaller than $e^{\pm m_s T}$, we know that the discrete model (2.11) will exhibit the two-time-scale property.

Calling h as the small parameter of the discrete-model, it follows that

$$\frac{e^{-m_f T}}{e^{\pm m_s T}} \sim h,$$

from which

$$T \cong \frac{1}{m_f \pm m_s} \ln\left(\frac{1}{h}\right) \qquad\qquad \ldots \text{(2.12)}$$

But from the singular perturbation theory, we have the approximate relationships

$$m_f \pm m_s = m_f, \text{ since } m_f \gg m_s \text{ , and}$$

$$m_f = g_{22}/\varepsilon$$

Using the above approximate relationships, (2.12) is written as

$$T = \frac{\epsilon}{g_{22}} \ln(\frac{1}{h}) \qquad \ldots (2.13a)$$

or

$$T = T_f \ln(\frac{1}{h}) \qquad \ldots (2.13b)$$

where $T_f (= \frac{1}{m_f} = \frac{\epsilon}{g_{22}})$ is the time constant of the fast mode of the continuous system.

It is important to note that in (2.13) ϵ and h are the singular perturbation parameters of the continuous model (2.10) and the corresponding discrete model (2.aa) respectively.

Case (b): Fast mode unstable

Let us consider the case when g_{22} is positive in (2.10) which corresponds to the unstable fast mode. The stability of the slow mode depends upon the other elements of the system matrix. Let the eigenvalues corresponding to the slow and fast modes be m_f amd $\pm m_s$ respectively. The two eigenvalues of the corresponding discrete model P_f and P_s are given by

$$P_f = e^{m_f T} \;;\; P_s = e^{\pm m_s T}$$

If T is chosen such the $e^{m_f T}$ is much larger than $e^{\pm m_s T}$, we know that the discrete model will exhibit the two-time-scale property. Calling h as the singular perturbation parameter of the discrete model and proceeding as in the previous case,

$$\frac{e^{m_f T}}{e^{\pm m_s T}} \simeq \frac{1}{h}, \text{ from which}$$

$$T = \frac{1}{(m_f \pm m_s)} \ln(\frac{1}{h})$$

Since $|m_f| \gg |m_s|$, $(m_f \pm m_s) \simeq m_f$

$$m_f \simeq g_{22}/\epsilon$$

Using the above approximate relationships,

$$T = \frac{\varepsilon}{g_{22}} \ln\left(\frac{1}{h}\right) \qquad \ldots \quad (2.14a)$$

or

$$T = T_f \ln\left(\frac{1}{h}\right) \qquad \ldots \quad (2.14b)$$

where T_f $(= \frac{1}{m_f} = \frac{\varepsilon}{g_{22}})$ is the time constant of the fast mode of the continuous system.

The equations (2.13) and (2.14) give the same value of T regardless of the sign of g_{22} in (2.10). But in case (a) P_f is much smaller than P_s. Equations (2.13b) and (2.14b) indicate that T must be chosen greater than T_f for getting a singularly perturbed discrete model.

Case study of a second order numerical example

Given the second order singularly perturbed differential equation

$$\begin{bmatrix} \dot{x}(t) \\ \dot{z}(t) \end{bmatrix} = \begin{bmatrix} g_{11} & g_{12} \\ \dfrac{g_{21}}{\varepsilon} & \dfrac{g_{22}}{\varepsilon} \end{bmatrix} \begin{bmatrix} x(t) \\ z(t) \end{bmatrix} \qquad \ldots \quad (2.15a)$$

its discrete model is

$$\begin{matrix} x(k+1) \\ z(k+1) \end{matrix} = \begin{matrix} a_{11} & a_{12} \\ a_{21} & a_{22} \end{matrix} \begin{matrix} x(k) \\ z(k) \end{matrix} \qquad \ldots \quad (2.15b)$$

where

$$a_{11} = \frac{(m_f - g_{22}/\varepsilon)e^{m_f T} - (m_s - g_{22}/\varepsilon)e^{m_s T}}{(m_f - m_s)}$$

$$a_{12} = \frac{g_{12}(e^{m_f T} - e^{m_s T})}{(m_f - m_s)}$$

$$a_{21} = \frac{(g_{12}/\varepsilon)(e^{m_f T} - e^{m_s T})}{m_f - m_s}$$

and

$$a_{22} = \frac{(m_f - g_{11})\, e^{m_f T} - (m_s - g_{11})\, e^{m_s T}}{(m_f - m_s)}$$

Case (a): Stable fast mode:

$g_{11} = 0$, $g_{12} = 1.0$, $g_{21} = -1.0$, and $g_{22} = -1.0$.

For $\varepsilon = 0.025$, 0.05, 0.075, 0.1 and 0.15, the discrete models are obtained for various discretizing intervals and the results are tabulated as shown in Table 2.1a with the following notation.

m_f and m_s are the eigenvalues of the continuous system corresponding to the fast and slow modes respectively.

P_f and P_s are the eigenvalues of the discrete model corresponding to the fast and slow modes respectively.

T is the discretizing interval.

T_f and T_s are the time constants of the continuous system corresponding to the fast and slow modes respectively.

a_{11}, a_{12}, a_{21}, and a_{22} are the elements of the system matrix of the discrete model.

A close examination of results shown in Table 2.1a indicate that

(i) If T is chosen as in (2.13), the resulting discrete model is in the singularly perturbed form and cast automatically in C-model (2.3). This is evident from the fact that the moduli of a_{11} and a_{21} are much larger than that of a_{12} and a_{22}.

(ii) An increase in T results in a higher ratio of P_s/P_f. For the numerical example considered, T = 1.5 to 2 times ε is sufficient to give singularly perturbed discrete model. For the values of ε less than the range considered (ε less than 0.025), T less than 1.5 ε is sufficient.

(iii) If T is chosen such that it is equal to $\varepsilon \ln(1/\varepsilon)$, P_s/P_f is nearly equal to m_f/m_s.

TABLE - 2.1a

(i) $\varepsilon = 0.025$
$m_f = -38.9737$, $m_s = -1.0263$, $T_f = 0.0257$ secs, $T_s = 0.9472$ secs, $m_f/m_s = 37.9737$

T/ε	a_{11}	a_{12}	a_{21}	a_{22}	P_s	P_f	P_s/P_f
1.0	0.9908	0.0157	-0.6295	0.3613	0.9747	0.3774	2.5823
1.5	0.9820	0.0192	-0.7699	0.2121	0.9622	0.2319	4.1497
2.0	0.9718	0.0213	-0.8512	0.1206	0.9500	0.1425	6.6683
2.5	0.9609	0.0224	-0.8963	0.0645	0.9379	0.0875	10.7157
$\ln(\frac{1}{\varepsilon})$	0.9336	0.0232	-0.9299	0.0036	0.9097	0.0275	33.0800

(ii) $\varepsilon = 0.05$
$m_f = -18.9443$, $m_s = -1.0263$, $T_f = 0.0528$ secs, $T_s = 0.9472$ secs, $m_f/m_s = 17.9443$

T/ε	a_{11}	a_{12}	a_{21}	a_{22}	P_s	P_f	P_s/P_f
1.0	0.9817	0.0313	-0.6270	0.3547	0.9486	0.3879	2.4459
1.5	0.9641	0.0381	-0.7629	0.2012	0.9239	0.2415	3.8253
2.0	0.9440	0.0419	-0.8379	0.1062	0.8998	0.1504	5.9826
2.5	0.9226	0.0438	-0.8751	0.0475	0.8764	0.0937	9.3565
$\ln(\frac{1}{\varepsilon})$	0.9007	0.0445	-0.8890	0.0116	0.8537	0.0586	14.5773

(iii) $\varepsilon = 0.075$
$m_f = -12.2444$, $m_s = -1.0889$, $T_f = 0.0817$ secs, $T_s = 0.9183$ secs, $m_f/m_s = 11.2444$

T/ε	a_{11}	a_{12}	a_{21}	a_{22}	P_s	P_f	P_s/P_f
1.0	0.9726	0.0468	-0.6244	0.3482	0.9216	0.3992	2.3086
1.5	0.9464	0.0567	-0.7560	0.1905	0.8847	0.2522	3.5078
2.0	0.9167	0.0618	-0.8247	0.0920	0.8493	0.1593	5.3302
2.5	0.8851	0.0641	-0.8547	0.0309	0.8153	0.1007	8.0983
$\ln(\frac{1}{\varepsilon})$	0.8793	0.0642	-0.8566	0.0227	0.8093	0.0927	8.7336

(iv) $\epsilon = 0.1$
$m_f = -8.8730$, $m_s = -1.1270$, $T_f = 0.1127$, $T_s = 0.8873$,
$m_f/m_s = 7.8730$

T/ϵ	a_{11}	a_{12}	a_{21}	a_{22}	P_s	P_f	P_s/P_f
1.0	0.9635	0.0622	-0.6218	0.3417	0.8934	0.4118	2.1697
1.5	0.9289	0.0749	-0.7491	0.1798	0.8445	0.2642	3.1960
2.0	0.8897	0.0812	-0.8116	0.0781	0.7982	0.1696	4.7077
2.5	0.8484	0.0834	-0.8335	0.0149	0.8633	0.0821	6.9344
$\ln(\frac{1}{\epsilon})$	0.8648	0.0829	-0.8286	0.0362	0.7714	0.1296	5.9511

(v) $\epsilon = 0.15$
$m_f = -5.4415$, $m_s = -1.2251$, $T_f = 0.1838$ secs,
$T_s = 0.8162$ secs, $m_f/m_s = 4.4415$

T/ϵ	a_{11}	a_{12}	a_{21}	a_{22}	P_s	P_f	P_s/P_f
1.0	0.9455	0.0925	-0.6167	0.3288	0.8321	0.4421	1.8822
1.5	0.8942	0.1103	-0.7354	0.1588	0.7591	0.2940	2.5823
2.0	0.8368	0.1179	-0.7858	0.0510	0.6924	0.1954	3.5435
2.5	0.7744	0.1190	-0.7932	-0.0667	0.6448	0.0660	9.7696
$\ln(\frac{1}{\epsilon})$	0.8489	0.1169	-0.7796	0.0693	0.7056	0.2126	3.3189

Case (b):   Unstable fast mode:

$g_{11} = 0$, $g_{12} = 1.0$, $g_{21} = -1.0$, and $g_{22} = 1.0$

The analysis of (2.15) is carried out with the above numerical values of G matrix as in Case (a) and the results are tabulated in Table 2.1b.

A close examination of the results in Table 2.1b leads to the same conclusions with respect to the choice of the discretizing interval T as in case (a). But in this case, the resulting discrete model is cast automatically in D-model (2.6). This is evident from the fact that the moduli of a_{21} and a_{12}.

Comments on the singularly perturbed discrete models of continuous systems

The above analysis leads us to an important conclusion that in order to get the discrete model in the singularly perturbed form, the discretizing interval T should be selected slightly greater than ε (the singular perturbation parameter of the continuous model). But we are aware that for an accurate discrete model, T must be chosen much smaller than the smallest time constant of the system [13]. On this basis if we select T much smaller than T_f (time constant of the fast mode) we get an extremely small value of T. For example in the case study of the second order system above, for $\varepsilon = 0.05$, T_f is very small. Knowing fully well that the fast mode is significant only in the "inner region" of the order of ε (inside the boundary layer), it is unnecessary to discretize the system for the entire region.

TABLE - 2.1b

(i) $\varepsilon = 0.025$
$m_f = 38.9737$; $m_s = 1.0263$; $T_f = 0.0257$ secs, $T_s = 0.9743$ secs,
$m_f/m_s = 37.9737$

T/E	a11	a_{12}	a_{21}	a_{22}	P_f	P_s	P_f/P_s
1.0	0.9821	0.0428	-1.7112	2.6933	2.6494	1.0260	2.5823
1.5	0.9507	0.0863	-3.4503	4.4010	4.3125	1.0392	4.1498
2.0	0.8913	0.1572	-6.2895	7.1808	7.0194	1.0527	6.6680
2.5	0.7861	0.2730	-10.9197	11.7958	11.4256	1.0662	10.7162
$\ln(\frac{1}{\varepsilon})$	0.1457	0.9291	-37.1644	37.3100	36.3564	1.0993	33.0724

(ii) $\varepsilon = 0.05$
$m_f = 18.9443$; $m_s = 1.0557$, $T_f = 0.0528$ secs, $T_s = 0.9472$ secs,
$m_f/m_s = 17.9443$

T/ε	a_{11}	a_{12}	a_{21}	a_{22}	P_f	P_s	P_f/P_s
1.0	0.9642	0.0852	-1.7042	2.6685	2.5785	1.0542	2.4459
1.5	0.9019	0.1710	-3.4191	4.3210	4.1405	1.0824	3.8253
2.0	0.7846	0.3095	-6.1910	6.9755	6.6488	1.1113	5.9829
2.5	0.5783	0.5330	-10.6608	11.2392	10.6764	1.1411	9.3562
$\ln(\frac{1}{\varepsilon})$	0.2325	0.8893	-17.7854	18.0179	17.0791	1.1714	14.5801

(iii) $\epsilon = 0.075$
$m_f = 12.2444$, $m_s = 1.0889$, $T_f = 0.0817$ secs, $T_s = 0.9183$ secs,
$_f/m_s = 11.2444$

T/ϵ	a_{11}	a_{12}	a_{21}	a_{22}	P_f	P_s	P_f/P_s
1.0	0.9465	0.1273	-1.6972	2.6437	2.5051	1.0851	2.3086
1.5	0.8536	0.2541	-3.3880	4.2416	3.9649	1.1303	3.5078
2.0	0.6798	0.4570	-6.0934	6.7732	6.2755	1.1774	5.3299
2.5	0.3767	0.7804	-10.4058	10.7824	9.9326	1.2265	8.0983
$\ln(\frac{1}{\epsilon})$	0.3029	0.8566	-11.4210	11.7234	10.7906	1.2356	8.7331

(iv) $\epsilon = 0.1$
$m_f = 8.8730$; $m_s = 1.1270$, $T_f = 0.1127$ secs, $T_s = .8873$ secs,
$m_f/m_s = 7.8730$.

T/ϵ	a_{11}	a_{12}	a_{21}	a_{22}	P_f	P_s	P_f/P_s
1.0	0.9288	0.1690	-1.6902	2.6191	2.4286	1.1193	2.1697
1.5	0.8058	0.3357	-3.3572	4.1630	3.7846	1.1842	3.1959
2.0	0.5770	0.5997	-5.9968	6.5738	5.8979	1.2528	4.7078
2.5	0.1810	1.0155	-10.0155	10.3357	9.1912	1.3254	6.9347
$\ln(\frac{1}{\epsilon})$	0.3621	0.8289	-8.2893	8.6514	7.7172	1.2963	5.9533

(v) $\epsilon = 0.15$
$m_f = 5.4415$; $m_s = 1.2251$, $T_f = 0.1838$ secs, $T_s = 0.8162$ secs,
$m_f/m_s = 4.4415$

T/ϵ	a_{11}	a_{12}	a_{21}	a_{22}	P_f	P_s	P_f/P_s
1/0	0.8937	0.2514	-1.6763	2.5700	2.2619	1.2018	1.8821
1.5	0.7117	0.4944	-3.2959	4.0076	3.4019	1.3174	2.5823
2.0	0.3772	0.8709	-5.8063	6.1834	5.1163	1.4442	3.5427
2.5	-0.1927	1.4495	-9.6635	9.4708	7.6948	1.5832	4.8603
$\ln(\frac{1}{\epsilon})$	0.4618	0.7798	-5.1986	5.6605	4.7051	1.4172	3.3200

with such a small T and no numerical advantage is gained by this
approach of solving the differential equation. On the other hand, if
we choose T on the basis of T_s , the time-constant of the slow mode,
there are certain advantages from the computation point of view.
For $\varepsilon = 0.05$, T $_s$ is equal to 0.9472 seconds, and we can select T equal
to 0.1 seconds which is nearly one-tenth of T_s . Thus it is concluded
that

   (i) If T is chosen on the basis of T_f, it will be very small and
no numerical advantage is gained by this approach. Moreover, the
discret model will not be in the singularly perturbed form.

   (ii) If T is chosen on the basis of T $_f$, the resulting discrete
model is good from the computational point of view. Also the model is
automatically brought to one of the standard models (C- or D-models)
amenable for singular perturbation analysis in the discrete version.
This discrete model can be considered as a fairly accurate version of
the original continuous system except in the inner region.

Effect of the fast mode

   If the effect of the fast mode is also of interest, a different
model with much smaller T on the basis of T_f is to be considered. We
have seen that this model will not be in the singularly perturbed form.
However since the fast mode is significant only in the inner region
with the thickness of the order of ε, it is enough if this model is
solved for this region only. Therefore solving the full-order equations for a few discrete intervals inside the boundary layer is not
difficult. Moreover, no separate efforts are needed to get the
discrete model as Euler's approximation is good enough for this small
value of T.

   Suppose we are interested in solving a two point boundary value
problem of a stiff differential equation via its discrete model. Given
the boundary conditions, say x(N) and z(0) for (2.8), the most important step in getting the solution is the determination of the missing
initial condition x(0). All numerical difficulties due to the stiffness and higher order are encountered chiefly in evaluation of x(0).
Once this is determined, the rest is a simple procedure to treat it
like an initial value problem. It is important to note that while
getting the solution of the discrete model (on the bisis of T_s) using

the singular perturbation method, x(0) is obtained upto the desired order of approximation taking advantage of the reduction of order and the removal of stiffness. Therefore for the solution inside the boundary layer both x(0) and z(0) are available and one can get the solution as in the initial value problem. Thus even for the boundary value problem, solution inside the boundary layer is easily obtained treating it as an initial value problem.

A numerical example (2.3) which highlights the various implications discussed above is provided in section 2.5 after the analysis of C-model is discussed.

General schemes of difference approximations and modelling errors

The general schemes of obtaining discrete models of continuous systems are by using forward difference, backward difference and central difference methods. The errors involved in obtaining the models using the above schemes depend on the discretizing interval T and are of the order of $0(T^2)$ for forward and backward differences, and $0(T^3)$ for central differences [13]. We have seen that for getting the discrete model in the singularly perturbed form, T should be chosen slightly greater than ε. Therefore, if the discrete models are obtained using these difference schemes, large errors due to the modelling itself will occur.

Similarly, in the suggested method of obtaining the models as in (2.9), if one adopts Euler's approximation writing

$$e^{AT} = I + AT$$

erros of the order of $0(T^2)$ will be introduced.

Keeping these points in view, in the evaluation of e^{AT}, higher order terms such as $\frac{A^2 T^2}{2!}$, $\frac{A^3 T^3}{3!}$, ... etc. are to be considered so that the errors due to the modelling are negligibly small in comparison with those that will be introduced due to the singular perturbation method.

2.4 Sampled-Data Control Systems

For the analysis of sampled-data control systems with zero order hold, if the singularly perturbed continuous part of the system is as in (2.8), its discrete model is given by (2.9). Thus the various aspects discussed in the above sectionare also valid for the case of sampled-data systems. However, the choice of sampling interval T in this case is purely governed by the significant time constants of the system as per Shanon's theorem. It is usually chosen less than about 1/4 times the significant time constant of the system [27].

In the case study of the second order example considered in the previous section, for $\epsilon = 0.05$, the two time constants T_s and T_f are 0.9472 seconds and 0.0528 secnds respectively. If we consider the smaller time constant T_f corresponding to the fast mode, the sampling interval has got to be less than 0.0132 second (1/4 of 0.0528 seconds), which is extremely small. On the other hand, if we consider the larger time constant T_s corresponding to the slow mode, T can be chosen less than 0.2368 seconds (1/4 of 0.9472 seconds). Because of the two-time-scale property of (2.8), we know that T_f is significant only in the "inner region" (inside the boundary layer) and T_s is significant in the "outer region". Under these circumstances, it is unnecessary and cumbersome to use such a small T less than 0.0132 seconds throughout the region.

Thus in the case of sampled-data control system such as the one described above, we can adopt multirate sampling, that is, a high rate of sampling with T less than $T_f/4$ in the outer region. We have to therefore use two different discrete models for mathematical anlysis such as in the design of controllers. We have seen in the previous section that the model based on T_s is automatically cast in the singularly perturbed C-model or D-model. If this model is used for the entire region, it has the disadvantage that the effect of the fast mode of the continuous system is folded up in the discretization process. Therefore if the effect of the fast mode is of interest, a different model on the basis of T_f is to be considered and analysed for the solution inside the boundary layer. We have seen in the previous section that this model though not in the singularly perturbed form can be easily handled.

In many sampled-data control systems, the effect of the fast modes can be ignored in the controller design for various reasons [27].

(i)   If there is a dead time or transportation lag larger than
ε in the system, the boundary layer is automatically folded up and the
effect of the fast mode is almost negligible.

(ii)   For the more practical case of the regulator design in which
we are only intereted in the steady state solution, the effect of the
fast mode is completely absent.

(iii)   The hold and the devices that follow it in a sampled-data
system are known to be low-pass filters and signals with very small
time constants are generally attenuated.

2.5 Singular Perturbation Analysis of State Space Models

In the following sections, the singular perturbation analysis of
the three state space models developed in the previous section is
carried out separately, and the various distinguishing features such as
order reduction, boundary layer, loss of boundary conditions, their
recovery etc., are clearly brought out.

Analysis of C-model [29,30]

Consider the (n+m)th order difference equation

$$
\begin{bmatrix} x(k+1) \\ z(k+1) \end{bmatrix} = \begin{bmatrix} A_{11} & hA_{12} \\ A_{21} & hA_{22} \end{bmatrix} \begin{bmatrix} x(k) \\ z(k) \end{bmatrix} + \begin{bmatrix} E \\ F \end{bmatrix} u(k) \qquad \ldots \ (2.16)
$$

x(k) is (nx1) state vector;

z(k) is (mx1) state vector;

u(k) is (rx1) control vector;

A_{11}, A_{12}, A_{21}, A_{22}, E and F are matrices of compatible dimensionality.

Letting h = 0 in (2.16), its degenerate form is

$$
x^{(0)}(k+1) = A_{11} x^{(0)}(k) + E u(k) \qquad \ldots \ (2.17a)
$$

$$
z^{(0)}(k+1) = A_{21} x^{(0)}(k) + F u(k) \qquad \ldots \ (2.17b)
$$

In the above two equations, only (2.17a) is the difference equation of order n. Equation (2.17b) is an algebraic relationship between $x^{(0)}(k)$ and $z^{(0)}(k)$. It means that once $x^{(0)}(k)$ is solved from (2.17a), the solution of $z^{(0)}(k)$ is automatically fixed from (2.17b). Thus since the order of the difference equation drops from (n+m) to n in the degeneration process, (2.16) is said to be in the singularly perturbed form.

In order to get a clear insight into the method and fix basic ideas, consider the second order equatin of the above model

$$\begin{bmatrix} x(k+1) \\ z(k+1) \end{bmatrix} = \begin{bmatrix} a_{11} & ha_{12} \\ a_{21} & ha_{22} \end{bmatrix} \begin{bmatrix} x(k) \\ z(k) \end{bmatrix} \qquad \ldots \quad (2.18)$$

Case (a): Initial value problem (IVP

Given $x(0)$ and $z(0)$, the solution of (2.18) is given by

$$\begin{bmatrix} x(k) \\ z(k) \end{bmatrix} = \begin{bmatrix} A \end{bmatrix}^k \begin{bmatrix} x(0) \\ z(0) \end{bmatrix} \; ; \; \text{where } A = \begin{bmatrix} a_{11} & h\,a_{12} \\ a_{21} & h\,a_{22} \end{bmatrix}$$

A^k is the state transition matrix of (2.18) which is obtained using Sylvester's formula [13] as

$$A^k = \frac{p_1^k \left[A - p_2 I \right] - p_2^k \left[A - p_1 I \right]}{(p_1 - p_2)} \qquad \ldots \quad (2.19)$$

where p_1 and p_2 are the eigenvalues of A given by

$$p_{1,2} = \frac{(a_{11}+ha_{22})}{2} \pm \frac{(a_{11}+ha_{22})}{2} \left[1 - \frac{4h(a_{11}a_{22} - a_{12}a_{21})}{(a_{11}+ha_{22})^2} \right]^{0.5} \ldots \quad (2.20)$$

Using (2.19) and (2.20), the solution of (2.18) is obtained. Expanding the terms under the square root of (2.20) binomially and ignoring terms with coefficients of h and higher powers of h in the solution thus obtained the zeroth order approximate solution of (2.18) is obtained as

$$x(k) = a_{11}{}^k x(0) + h^{k+1} \left[-\frac{a_{12}}{a_{11}} \{z(0) - \frac{a_{21}}{a_{11}} x(0)\} \{\frac{a_{11}a_{22} - a_{12}a_{21}}{a_{11}}\}^k \right]$$

$$\cdots \quad (2.21a)$$

$$z(k) = a_{21} a_{11}{}^{k-1} x(0) + h^k \left[\{z(0) - \frac{a_{21}}{a_{11}} x(0)\} \{\frac{a_{11}a_{22} - a_{12}a_{21}}{a_{11}}\}^k \right] \cdots (2.21b)$$

In (2.21a), the terms associated with h^{k+1} are retained for the reasons that will become apparent later.

The degenerate form of (2.18) is

$$\begin{bmatrix} x^{(0)}(k+1) \\ z^{(0)}(k+1) \end{bmatrix} = \begin{bmatrix} a_{11} \\ a_{21} \end{bmatrix} x^{(0)}(k)$$

If the above equation is solved with $x^{(0)}(0) = x(0)$, the solution is

$$x^{(0)}(k) = a^k{}_{11} x(0) \qquad\qquad \cdots \quad (2.22a)$$

$$z^{(0)}(k) = a_{21} a_{11}^{k-1} x(0) \qquad\qquad \cdots \quad (2.22b)$$

Note that the solutions (2.22a) and (2.22b) are the same as the terms not associated with h^k and h^{k+1} in (2.21a) and (2.21b) respectively.

A close examination of (2.21) and (2.22) reveals that

(i) As h tends to zero, the solution of $z(k)$ fails to uniformly converge to the degenerate solution $z^{(0)}(k)$ at $k = 0$, leading to the formation of the boundary layer at $k = 0$.

(ii) Once $x^{(0)}(k)$ is solved as in (2.22), the solution of $z^{(0)}(k)$ is automatically fixed and $z^{(0)}(0) \neq z(0)$.

(iii) The lost boundary condition can be readmitted in the form of boundary layer correction series. The structure of (2.21) suggests the transformations for the correction series as

$$v(k) = x(k)/h^{k+1}; \; w(k) = z(k)/h^k \qquad\qquad \cdots \quad (2.23)$$

The recovery of the lost initial condition is demonstrated as follows.

Using the above transformation (2.23) in (2.18) and dividing throughout with h^{k+1}, the equation of the correction series is obtained as

$$\begin{bmatrix} hv(k+1) \\ w(k+1) \end{bmatrix} \hat{=} \begin{bmatrix} a_{11} & a_{12} \\ a_{21} & a_{22} \end{bmatrix} \begin{bmatrix} v(k) \\ w(k) \end{bmatrix} \qquad \dots \quad (2.24)$$

The degenerate for of the above equation can be written as

$$w^{(0)}(k+1) = \left[\frac{a_{11}a_{22} - a_{12}a_{21}}{a_{11}} \right] w^{(0)}(k)$$

$$v^{(0)}(k) = - \frac{a_{12}}{a_{11}} w^{(0)}(k)$$

Choosing $w^{(0)}(0) = z(0) - z^{(0)}(0)$, and writing $z^{(0)}(0) = \frac{a_{21}}{a_{11}} x(0)$ from (2.22b) the solution of the above degenerate equation is

$$w^{(0)}(k) = \left[z(0) - \frac{a_{21}}{a_{11}} x(0) \right] \left[\frac{a_{11} a_{22} - a_{12} a_{21}}{a_{11}} \right]^k$$

$$v^{(0)}(k) = - \frac{a_{12}}{a_{11}} \left[z(0) - \frac{a_{21}}{a_{11}} x(0) \right] \left[\frac{a_{11} a_{22} - a_{12} a_{21}}{a_{11}} \right]^k$$

Note that the solutions $v^{(0)}(k)$ and $w^{(0)}(k)$ obtained above are the same terms in (2.21a) and (2.21b) associated with h^{k+1} and h^k respectively. This clearly demonstrates the recovery of the lost boundary condition. The total series solution for the zeroth order approximation is

$$x(k) = x^{(0)}(k) + h^{k+1} v^{(0)}(k)$$

$$z(k) = z^{(0)}(k) + h^k w^{(0)}(k)$$

which satisfy both the given initial conditions exactly.

Case (b): Boundary value problem (BVP)

For the boundary value problem of (2.18), given $x(N)$ and $z(0)$, the zeroth order solution is obtained from the zeroth order solution of IVP (2.21) as below.

Letting $k = N$ in (2.21a)

$$x(0) = a_{11}^{-N} x(N) + 0(h)$$

Using the above value of $x(0)$ in (2.21a, b), the zeroth order solution of BVP is

$$x(k) = a_{11}^{k-N} x(N) +$$

$$h^{k+1} \left[- \frac{a_{12}}{a_{11}} \{ z(0) - \frac{a_{21}}{a_{11}} a_{11}^{-N} x(N) \} \{ \frac{a_{11} a_{22} - a_{12} a_{21}}{a_{11}} \}^k \right] \quad \ldots \quad (2.25a)$$

$$z(k) = a_{21} a_{11}^{k-N-1} x(N) +$$

$$h^k \left[(z(0) - \frac{a_{21}}{a_{11}} a_{11}^{-N} x(N) \} \{ \frac{a_{11} a_{22} - a_{12} a_{21}}{a_{11}} \}^k \right] \quad \ldots \quad (2.25b)$$

If the degenerate form of (2.18), that is,

$$\begin{bmatrix} x^{(0)}(k+1) \\ z^{(0)}(k+1) \end{bmatrix} = \begin{bmatrix} a_{11} \\ a_{21} \end{bmatrix} x^{(0)}(k)$$

is solved with $x^{(0)}(N) = x(N)$, the solutions are obtained as

$$x^{(0)}(k) = a_{11}^{k-N} x(N) \quad \ldots \quad (2.26a)$$

$$z^{(0)}(k) = a_{21} a_{11}^{k-N-1} x(N) \quad \ldots \quad (2.26b)$$

Note that the above solutions $x^{(0)}(k)$ and $z^{(0)}(k)$ are the same terms not associated with h^{k+1} and h^k in (2.25a) and (2.25b) respectively.

A close examination of (2.25) and (2.26) reveals

(i) As h tends to zero, the solution of z(k) fails to uniformly converge to the degenerate solution $z^{(0)}(k)$ at k = 0, <u>leading to the formation of the boundary layer at k = 0.</u>

(ii) In the process of degeneration z(k) loses its initial conditions z(0), and

$$z(0) \neq z^{(0)}(0)$$

(iii) The transformations required for the correction series are as in (2.23), same as in IVP. Consequently, the equation for the correction series is (2.24), same as in IVP.

As in the case of IVP, the recovery of the lost initial condition can be demonstrated as below. The degenerate form of the correction series is

$$w^{(0)}(k+1) = \left[\frac{a_{11} \, a_{22} - a_{12} \, a_{21}}{a_{11}} \right] w^{(0)}(k)$$

$$v^{(0)}(k) = - \frac{a_{12}}{a_{11}} w^{(0)}(k)$$

Choosing $w^{(0)}(0) = z(0) - z^{(0)}(0)$, and writing $z^{(0)}(0) = (a_{21} / a_{11}) a_{11}^{-N} x(N)$ from (2.26b), the resulting solutions $v^{(0)}(k)$ and $w^{(0)}(k)$ are found to be same as the terms associated with h^{k+1} and h^k in (2.25a) and (2.25b) respectively. Thus the total series solution for zeroth order zpproximation is written as

$$x(k) = x^{(0)}(k) + h^{k+1} v^{(0)}(k)$$

$$z(k) = z^{(0)}(k) + h^k w^{(0)}(k)$$

which satisfies both the boundary conditions.

Analysis of higher order model

Consider the (n+m)th order model (2.16). Suppressing h the degenerate equation obtained (2.17) is of order n. Based on the results of the second order system obtained above, one can conclude that the degenerate equation (2.17) satisfies the n initial conditions x(0) and loses the remaining m initial conditions z(0). This is clearly illustrated by means of a block diagram in Fig. 2.1.

Using the same transformations (2.23), a correction series is incorporated to recover the lost initial conditions. The total series solution of (2.16) is therefore written as

$$x(k) \quad = \quad x_t(k) + h^{k+1} v(k) \qquad \qquad \dots \quad (2.27a)$$

$$z(k) \quad = \quad z_t(k) + h^k w(k) \qquad \qquad \dots \quad (2.27b)$$

where $x_t(k)$ and $z_t(k)$ refer to the outer series and v(k) and w(k) refer to the correction series respectively.

Substituting (2.27) in (2.26), and separating the outer series and correction series terms, equations for the outer series and correction series are obtained as shown.

Outer Series:

$$\begin{bmatrix} x_t(k+1) \\ z_t(k+1) \end{bmatrix} = \begin{bmatrix} A_{11} & hA_{12} \\ A_{21} & hA_{22} \end{bmatrix} \begin{bmatrix} x_t(k) \\ z_t(k) \end{bmatrix} + \begin{bmatrix} E \\ F \end{bmatrix} u(k) \qquad \dots \quad (2.28)$$

Correction Series:

$$\begin{bmatrix} h^{k+2} \, v(k+1) \\ h^{k+1} \, w(k+1) \end{bmatrix} = \begin{bmatrix} A_{11} & A_{12} \\ A_{21} & A_{22} \end{bmatrix} \begin{bmatrix} h^{k+1} \, v(k) \\ h^{k+1} \, w(k) \end{bmatrix}$$

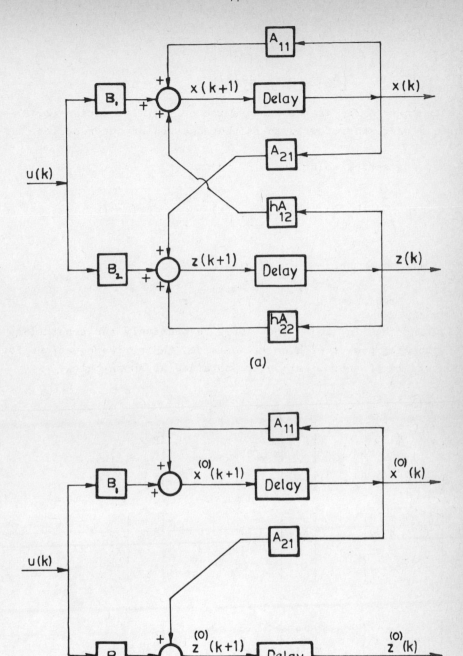

Fig. 2.1(a) Original system   (b) Degenerate system

or

$$\begin{bmatrix} hv(k+1) \\ w(k+1) \end{bmatrix} = \begin{bmatrix} A_{11} & A_{12} \\ A_{21} & A_{22} \end{bmatrix} \begin{bmatrix} v(k) \\ w(k) \end{bmatrix} \qquad \ldots \ (2.29)$$

Note that (2.28) is the same as the originial equation (2.16), which is henceforth referred to as the equation of outer series.

Assuming series solutions

$$x(k) = \sum_{r=0}^{\infty} h^r x^{(r)}(k) \quad ; \quad z(k) = \sum_{r=0}^{\infty} h^r z^{(r)}(k),$$

and

$$v(k) = \sum_{r=0}^{\infty} h^r v^{(r)}(k) \quad ; \quad w(k) = \sum_{r=0}^{\infty} h^r w^{(r)}(k)$$

substituting them in (2.16) and (2.29) respectively and separating terms with like powers of h on either side, the series equations for various orders of approximation are obtained as shown below.

Outer Series:

$$h^0 : \begin{bmatrix} x^{(0)}(k+1) \\ z^{(0)}(k+1) \end{bmatrix} = \begin{bmatrix} A_{11} \\ A_{21} \end{bmatrix} x^{(0)}(k) + \begin{bmatrix} E \\ F \end{bmatrix} u(k)$$

$$h^1 : \begin{bmatrix} x^{(1)}(k+1) \\ z^{(1)}(k+1) \end{bmatrix} = \begin{bmatrix} A_{11} \\ A_{21} \end{bmatrix} x^{(1)}(k) + \begin{bmatrix} A_{12} \\ A_{22} \end{bmatrix} z_0^{(0)}(k)$$

$$\ldots \quad \ldots\ldots\ldots \quad \ldots\ldots\ldots\ldots\ldots\ldots\ldots\ldots$$

$$h^j : \begin{bmatrix} x^{(j)}(k+1) \\ z^{(j)}(k+1) \end{bmatrix} = \begin{bmatrix} A_{11} \\ A_{21} \end{bmatrix} x^{(j)}(k) + \begin{bmatrix} A_{12} \\ A_{22} \end{bmatrix} z^{(j-1)}(k) \qquad \ldots \ (2.30)$$

Correction Series:

$$h^0 : \begin{bmatrix} 0 \\ w^{(0)}(k+1) \end{bmatrix} = \begin{bmatrix} A_{11} & A_{12} \\ A_{21} & A_{22} \end{bmatrix} \begin{bmatrix} v^{(0)}(k) \\ w^{(0)}(k) \end{bmatrix}$$

$$h^1 : \begin{bmatrix} v^{(0)}(k+1) \\ w^{(1)}(k+1) \end{bmatrix} = \begin{bmatrix} A_{11} & A_{12} \\ A_{21} & A_{22} \end{bmatrix} \begin{bmatrix} v^{(1)}(k) \\ w^{(1)}(k) \end{bmatrix}$$

$$\cdots \quad \cdots \cdots \quad \cdots \cdots \cdots \cdots$$

$$h^j : \begin{bmatrix} v^{(j-1)}(k+1) \\ w^{(j)}(k+1) \end{bmatrix} = \begin{bmatrix} A_{11} & A_{12} \\ A_{21} & A_{22} \end{bmatrix} \begin{bmatrix} v^{(j)}(k) \\ w^{(j)}(k) \end{bmatrix} \qquad \cdots \quad (2.31)$$

It is important to note that al the outer series equations are of order n while the correction series equations are of order m. It is essential that A_{11} is nonsingular for the above equations to be able to be solved.

The total series solution can be written as the sum of the solution of outer series and correction series as

$$x(k) = \sum_{r=0}^{\infty} [x^{(r)}(k) + h^{k+1}v^{(r)}(k)] \, h^r \qquad \cdots \quad (2.32a)$$

$$z(k) = \sum_{r=0}^{\infty} [z^{(r)}(k) + h^k w^{(r)}(k)] \, h^r \qquad \cdots \quad (2.32b)$$

The coefficients h^{k+1} and h^k in (2.32) indicate that the correction series terms v(k) and w(k) are to be solved for a few values of k near k = 0, depending on the order of approximation desired.

For the _initial value problem_ given $x(0)$ and $z(0)$ the initial conditions required to solve the series equations (2.30) and (2.31) are obtained from the total series solution (2.32) based on the fact that the two series together satisfy all the given conditions exactly. They are furnished below.

$$x^{(0)}(0) = x(0) \qquad ; \quad w^{(0)}(0) = z(0) - z^{(0)}(0)$$

$$x^{(1)}(0) = -v^{(0)}(0); \quad w^{(1)}(0) = -z^{(1)}(0)$$

$$\cdots \cdots \qquad \cdots \cdots \qquad \cdots \cdots \qquad \cdots \cdots$$

$$x^{(j)}(0) = -v^{(j-1)}(0) \qquad ; \quad w^{(j)}(0) = - z^{(j)}(0)$$

For the _boundary value problem_ of (2.16), given $x(N)$ and $z(0)$, the equations for outer series, correction series, the total series solution etc. are the same as in the initial value problem. The boundary conditions needed to solve the series equations which are also the same as in IVP, that is, (2.30) and (2.31) are however different. They are obtained from (2.32) as below

$$x^{(0)}(N) = x(N) \quad ; \quad w^{(0)}(0) = z^{(0)} - z^{(0)}(0)$$

$$x^{(1)}(N) = 0 \qquad ; \quad w^{(1)}(0) = -z^{(1)}(0)$$

$$\cdots \cdots \qquad \cdots \cdots \qquad \cdots \cdots \qquad \cdots \cdots \cdots$$

$$x^{(j)}(N) = 0 \qquad ; \quad w^{(j)}(0) = -z^{(j)}(0).$$

Example 2.1: C-model

Consider the fifth order example [25]

$$
\begin{bmatrix}
x_1(k+1) \\
x_2(k+1) \\
z_1(k+1) \\
z_2(k+1) \\
z_3(k+1)
\end{bmatrix}
=
\begin{bmatrix}
0.9014 & 0.1179 & 0.2100h & 0.0668h & 0.08416h \\
-0.0196 & 0.8743 & 0.0000h & 0.1000h & 0.11736h \\
-0.0071 & 0.7342 & 0.8070h & 0.0520h & 0.08427h \\
-0.75 & -0.0557 & -0.1280h & 0.77428h & -0.05630h \\
-0.306 & -0.01694 & -0.044h & 0.57112h & 0.05287h
\end{bmatrix}
\begin{bmatrix}
x_1(k) \\
x_2(k) \\
z_1(k) \\
z_2(k) \\
z_3(k)
\end{bmatrix}
$$

with the initial conditions

$$
\begin{bmatrix}
x_1(0) \\
x_2(0)
\end{bmatrix}
=
\begin{bmatrix}
1.0 \\
-0.8
\end{bmatrix}
;
\qquad
\begin{bmatrix}
z_1(0) \\
z_2(0) \\
z_3(0)
\end{bmatrix}
=
\begin{bmatrix}
0.5 \\
0.2 \\
0.6
\end{bmatrix}
; \text{ and } h = 0.25
$$

The eigenvalues of the above model are

$$p_{1,2} = 0.8777 \pm j\, 0.1054 \; ; \; p_3 = 0.0179 \; ;$$

$$p_{4,5} = 0.2055 \pm j\, 0.0236$$

The eigenvalues of the degenerate model (letting h = 0 in the above model) are

$$p_{1,2} = 0.8879 \pm j\, 0.0461$$

Fig. 2.2 clearly demonstrates the effect of the degeneration of the model and the formation of the boundary layer at k = 0.

Using the method developed above, the solutions of series equations are obtained for degenerate, zeroth and first order approximations. The results are plotted in Fig. 2.3. The exact and series solution coincide for a few values of k near k = 0 for the reason given in Example 1.1.

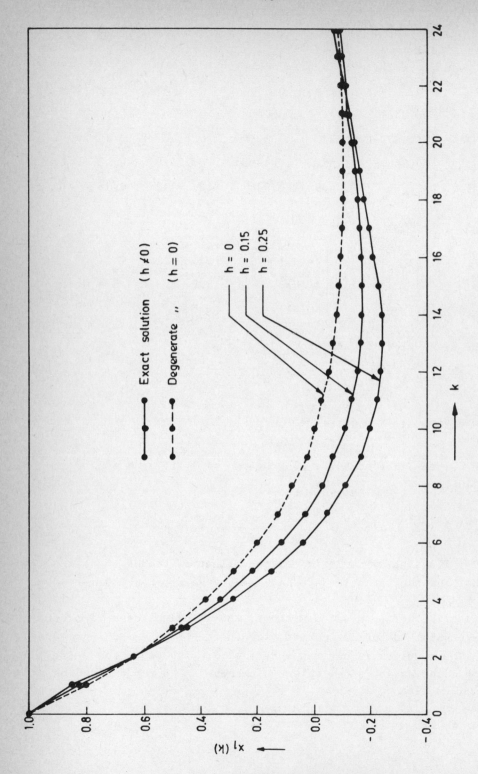

Fig. 2.2(a) Degeneration of $x_1(k)$ of Example 2.1

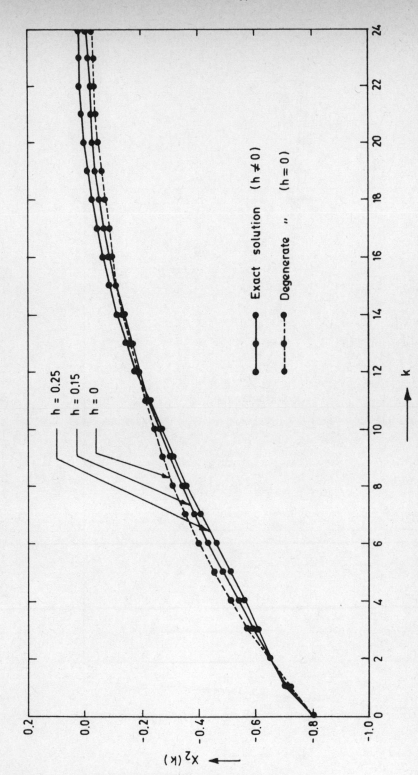

Fig. 2.2(b) Degeneration of $x_2(k)$ of Example 2.1

78

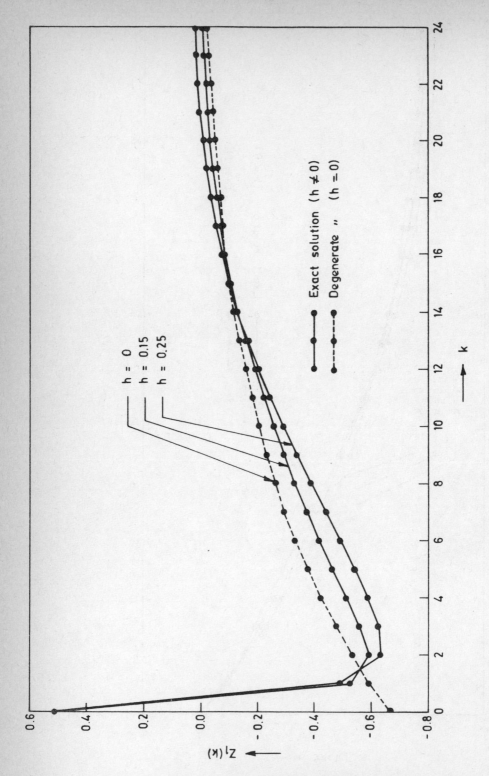

Fig. 2.2(c) Degeneration of $z_1(k)$ of Example 2.1

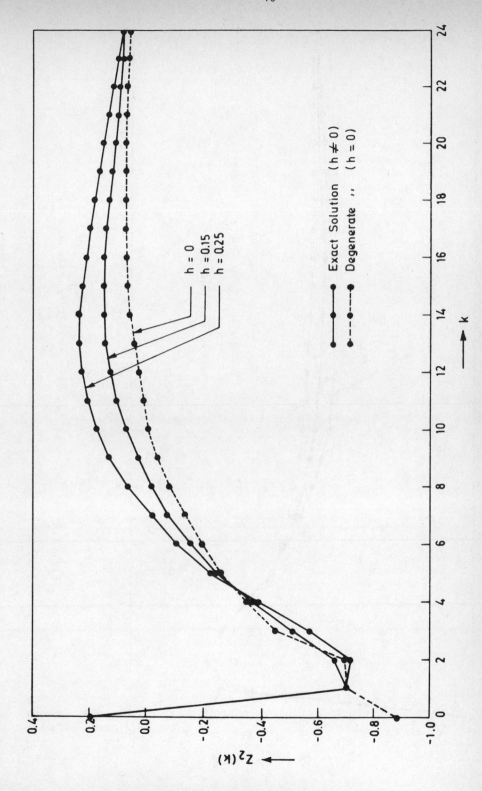

Fig. 2.2(d) Degeneration of $z_2(k)$ of Example 2.1

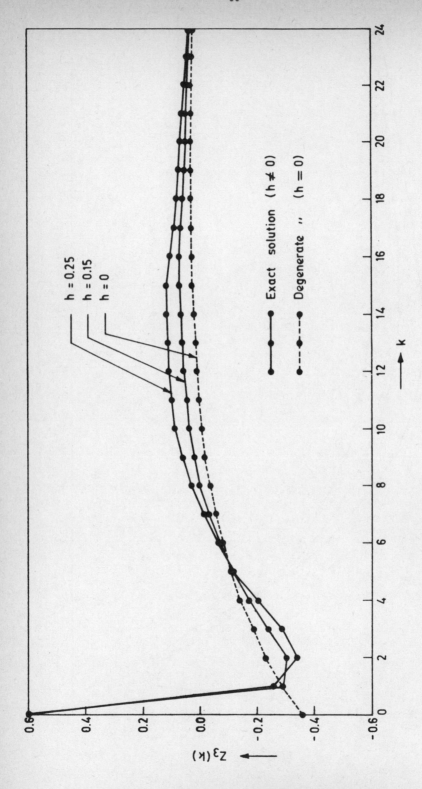

Fig. 2.2(e) Degeneration of $z_3(k)$ of Example 2.1

Example 2.2: C-model

Consider the fourth order two point boundary value problem [31]

$$
\begin{bmatrix} x_1(k+1) \\ x_2(k+1) \\ z_1(k+1) \\ z_2(k+1) \end{bmatrix} = \begin{bmatrix} 0 & 1 & 0 & 0 \\ -1.25 & -1.5 & -0.8h & -h \\ 0 & 0 & 0 & h \\ 1 & 0 & 0 & 0 \end{bmatrix} \begin{bmatrix} x_1(k) \\ x_2(k) \\ z_1(k) \\ z_2(k) \end{bmatrix} + \begin{bmatrix} 0 \\ 1 \\ 0 \\ 0 \end{bmatrix} u(k)
$$

with the boundary conditions

$$
\begin{bmatrix} x_1(10) \\ x_2(10) \end{bmatrix} = \begin{bmatrix} 0.6 \\ 0.5 \end{bmatrix} \quad ; \quad \begin{bmatrix} z_1(0) \\ z_2(0) \end{bmatrix} = \begin{bmatrix} 10.0 \\ 8.0 \end{bmatrix}
$$

h = 0.15 and u(k) is a unit step function.

The eigenvalues of the above model are

$$P_{1,2} = -0.6907 \pm j\,0.7696 \; ; \; P_{3,4} = -0.0593 \pm j\,0.1154$$

The eigenvalues of the degenerate model are

$$P_{1,2} = -0.75 \pm j\,0.8292$$

Using the method developed above, the series equations are solved for degenerate, zeroth, first and second order approximations. The results are compared with the exact solution in Table 2.2 which clearly illustrates that as the order of aproximation is increased, the series solutions approach the exact solution.
Note: The exact solution of the above two point boundary value problem can be found using either the method of complementary functions or the method of adjoints [44]. Both the methods are known as the shooting methods in which the first and the most important step is the evaluation of the missing initial conditions. Once this is done, the rest is a straightforward procedure to treat it as an initial value problem.

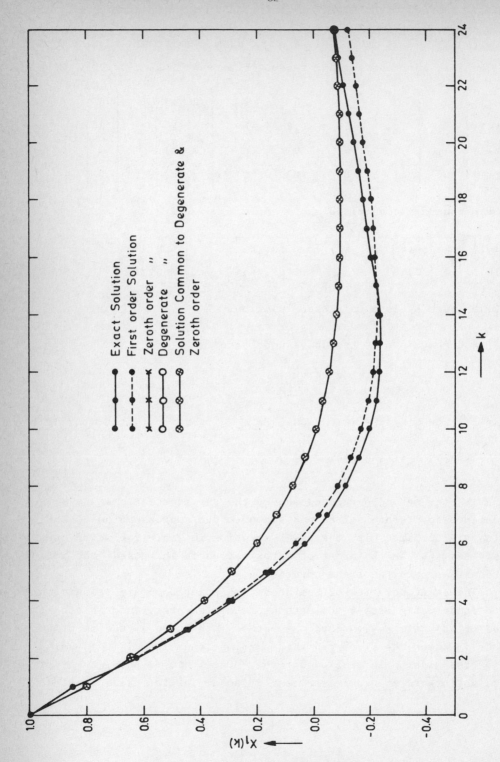

Fig. 2.3(a) Exact and approximate solutions of $x_1(k)$ of Example 2.1

83

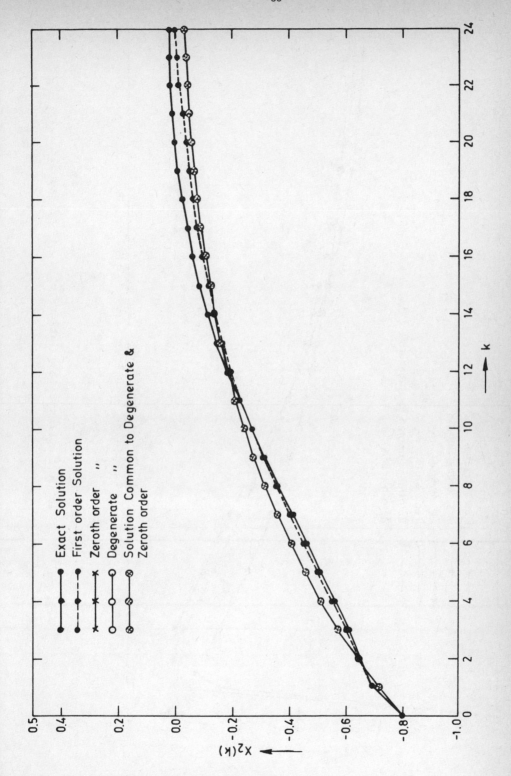

Fig. 2.3(b) Exact and approximate solutions of $x_2(k)$ of Example 2.1

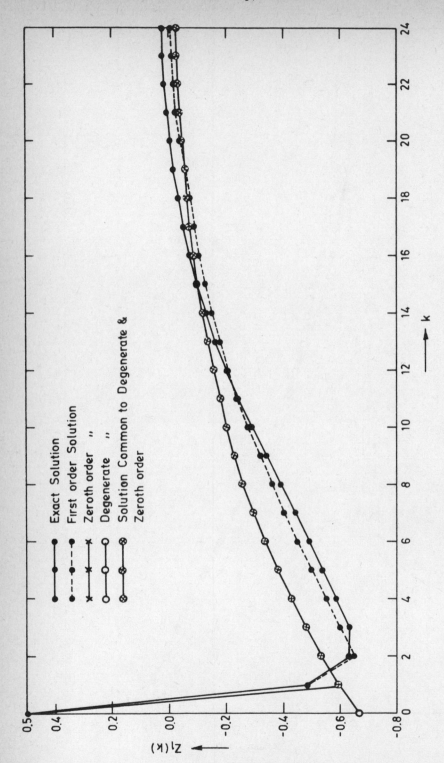

Fig. 2.3(c)   Exact and approximate solutions of $z_1(k)$ of Example 2.1

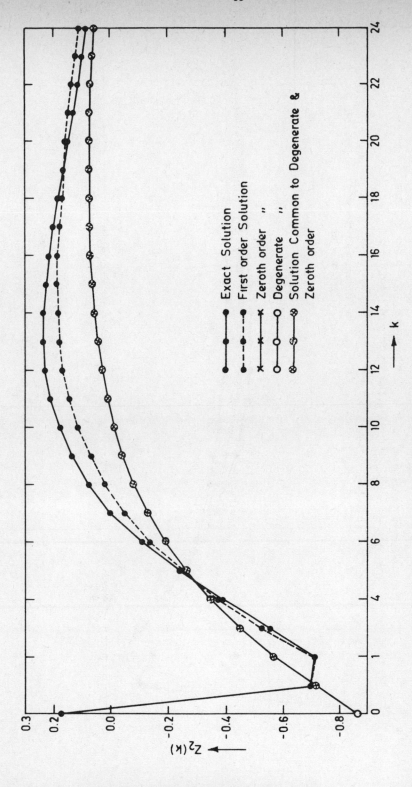

Fig. 2.3(d) Exact and approximate solutions of $z_2(k)$ of Example 2.1

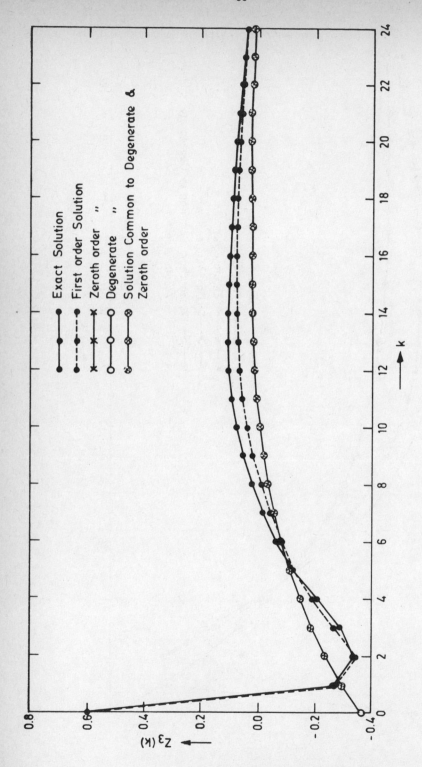

Fig. 2.3(e) Exact and approximate solutions of $z_3(k)$ of Example 2.1

In the method of complementary functions which is used in this example, the determination of $x_1(0)$ and $x_2(0)$ ultimately boils down to the solution of a pair of linear equations. But since the original system model possesses widely separated eigenvalues due to its singularly perturbed structure, these two equations become almost linearly dependent and it is impossible to determine $x_1(0)$ and $x_2(0)$ with acceptable accuracy. Under these circumstances, one has to resort to special algorithms such as Gram-Schmidt process and Conte's method [44] to ensure the linear independence of the matrix coefficients of the linear equations developed by the method of complementary functions.

This procedure for obtaining the exact solution is used for all the two-point boundary value problems in this monograph.

Example 2.3

Consider the boundary value problem of the continuous system described by the differential equation

$$
\begin{bmatrix} \dot{x}(t) \\ \varepsilon\dot{z}(t) \end{bmatrix} = \begin{bmatrix} -1 & 1 \\ -1 & -1 \end{bmatrix} \begin{bmatrix} x(t) \\ z(t) \end{bmatrix}
$$

with $x(1) = 2.0$ and $z(0) = -10.0$ as the boundary conditions and $\varepsilon = 0.05.$

The eigenvalues of the above system matrix are

$$m_f = -18.8815 \; ; \; m_s = -2.1185$$

Solution outside the boundary layer

Choosing the discretizing interval $T = 2\varepsilon = 0.1$, the discrete model is obtained as in (2.15), which is automatically in C-model as shown below.

TABLE - 2.2a

Comparison of approximate and exact solutions of x_1 (x)

of Example 2.2

k	Degenerate solution	Zeroth order solution	1st order solution	2nd order solution	Exact solution
0	0.3818	0.3818	-1.4321	-1.5272	-1.5117
1	0.0248	0.0248	-0.1535	-0.1109	-0.1510
2	0.4855	0.4855	0.6204	0.6753	0.7161
3	0.2407	0.2407	0.2040	0.1964	0.1973
4	0.0321	0.0321	-0.0852	-0.1226	-0.1413
5	0.6510	0.6510	0.8000	0.8449	0.8606
6	-0.0166	-0.0166	-0.1296	-0.1535	-0.1567
7	0.2112	0.2112	0.1895	0.1825	0.1770
8	0.7040	0.7040	0.7800	0.7965	0.8038
9	-0.3200	-0.3200	-0.4045	-0.4166	-0.4190
10	0.6000	0.6000	0.6000	0.6000	0.6000

TABLE - 2.2b

Comparison of approximate and exact solutions of x_2(k)

of Example 2.2

k	Degenerate solution	Zeroth order solution	1st order solution	2nd order solution	Exact solution
0	0.0248	0.0248	-0.1535	-0.1109	-0.1510
1	0.4855	0.4855	0.6204	0.6754	0.7161
2	0.2407	0.2407	0.2040	0.1964	0.1973
3	0.0321	0.0321	-0.0852	-0.1226	-0.1413
4	0.6510	0.6510	0.8000	0.8449	0.8606
5	-0.0166	-0.0166	-0.1296	-0.1535	-0.1567
6	0.2112	0.2112	0.1895	0.1825	0.1770
7	0.7040	0.7040	0.7800	0.7975	0.8038
8	-0.3200	-0.3200	-0.4045	-0.4166	-0.4190
9	0.6000	0.6000	0.6000	0.6000	0.6000
10	0.5000	0.5000	0.5000	0.5000	0.5000

89

TABLE - 2.2c

Comparison of approximate and exact solutions of $z_1(k)$

of Example 2.2

k	Degenerate solution	Zeroth order solution	1st order solution	2nd order solution	Exact solution
0	0.0000	10.0000	10.0000	10.0000	10.0000
1	0.0000	0.0000	1.2000	1.2000	1.2000
2	0.0000	0.0000	0.0573	1.2148	-0.2265
3	0.0000	0.0000	0.0037	-0.0230	-0.0227
4	0.0000	0.0000	0.0728	0.0931	0.1074
5	0.0000	0.0000	0.0361	0.0306	0.0296
6	0.0000	0.0000	0.0048	-0.0128	-0.0212
7	0.0000	0.0000	0.0977	0.1200	0.1291
8	0.0000	0.0000	-0.0025	-0.0194	-0.0235
9	0.0000	0.0000	0.0317	0.0284	0.0266
10	0.0000	0.0000	0.1056	0.1170	0.1206

Note: The degenerate and zeroth order solutions are the same except at k = 0

TABLE - 2.2d

Comparison of approximate and exact solution of $z_2(k)$

of Example 2.2

k	Degenerate solution	Zeroth order solution	1st order solution	2nd order solution	Exact solution
0	-0.4780	8.0000	8.0000	8.0000	8.0000
1	0.3818	0.3818	-1.4321	-1.5272	-1.5117
2	0.0248	0.0248	-0.1535	-0.1109	-0.1510
3	0.4855	0.4855	0.6204	0.6754	0.7161
4	0.2407	0.2407	0.2040	0.1964	0.1973
5	0.0321	0.0321	-0.0852	-0.1226	-0.1413
6	0.6510	0.6510	0.8000	0.8449	0.8606
7	-0.0166	-0.0166	-0.1296	-0.1535	-0.1567
8	0.2112	0.2112	0.1895	0.1825	0.1770
9	0.7040	0.7040	0.7800	0.7975	0.8038
10	-0.3200	-0.3200	-0.4045	-0.4166	-0.4190

Note: The degenerate and zeroth order solutions are the same except at k = 0

$$\begin{bmatrix} x(k+1) \\ z(k+1) \end{bmatrix} = \begin{bmatrix} 0.85297 & 0.3139h \\ -0.78475 & 0.85972h \end{bmatrix} \begin{bmatrix} x(k) \\ z(k) \end{bmatrix}$$

$x(10) = 2.0$; $z(0) = -10.0$ and $h = 0.125$

The eigenvalues of the above discrete model are

$p_f = 0.80908$; $p_s = 0.15135$

Using the singular perturbation method developed for C-model, the solution of the above discrete model is obtained upto second order approximation and the results are compared with the exact solution of the differential equation in Table 2.3a and Fig. 2.4.

Note: In the solution obtained as above, the effect of the fast mode of the continuous system which is significant inside the boundary layer is folded up between $k = 0$ and $k = 1$.

Solution inside the boundary layer

In order to take the fast mode of the continuous system also into account, a different discrete model with $T = \varepsilon/5 = 0.01$ secs is obtained using Euler's approximation as below.

$$\begin{bmatrix} x(k+1) \\ z(k+1) \end{bmatrix} = \begin{bmatrix} 0.9891 & 0.0090 \\ -0.1803 & 0.8179 \end{bmatrix} \begin{bmatrix} x(k) \\ z(k) \end{bmatrix}$$

The eigenvalues of the above model are

$p_1 = 0.9790$; $p_2 = 0.8280$

Evidently this model is not in the singularly perturbed form. Hence as explained in Sec. 2.2 this model is tackled as follows.

In obtaining the solution of the discrete model (with $T = 0.1$) the missing initial condition $x(0)$ is obtained for the degenerate, zeroth, first and second order approximations. Using the value of $x(0)$ thus obtained and the given $z(0)$, the full order discrete model (with $T = 0.01$) is solved as an initial value problem. Since the fast mode of the continuous system is significant nearly upto 0.05 secs (equal to the small parameter ε), it is enough if this discrete model is solved upto $k = 5$. Depending on the approximation used for $x(0)$, the resulting solutions may be called as the degenerate, zeroth, first and second order approximate solutions.

The results obtained are compared with the exact solution of the differential equation in Table 2.3b.

2.6 Analysis of R-Model [32]

The $(n+m)$th order singularly perturbed R-model is reproduced from (2.4b) as

$$\begin{bmatrix} x(k+1) \\ z(k+1) \end{bmatrix} = \begin{bmatrix} A_{11} & A_{12} \\ hA_{21} & hA_{22} \end{bmatrix} \begin{bmatrix} x(k) \\ z(k) \end{bmatrix} + \begin{bmatrix} E \\ hF \end{bmatrix} u(k) \qquad \ldots \text{(2.33a)}$$

Letting $h = 0$ in (2.33), the degenerate model is

$$x^{(0)}(k+1) = A_{11}x^{(0)}(k) + A_{12}z^{(0)}(k) + E u(k)$$

$$z^{(0)}(k+1) = 0$$

In the above two equations, the second equation has a trivial solution. In view of this, the degenerate equations are written as

$$x^{(0)}(k+1) = A_{11}x^{(0)}(k) + E u(k) \qquad \ldots \text{(2.33b)}$$

$$z^{(0)}(k+1) = 0 \qquad \ldots \text{(2.33c)}$$

From (2.33a) and (2.33b), it follows that the order of the equations drops from $(n+m)$ to n in the degeneration process.

TABLE - 2.3a

Comparison of approximate and exact solutions of Example 2.3
(outside the boundary layer)

x(t)/z(t)	Degenerate solution	Zeroth order solution	1st order solution	2nd order solution	Exact solution
x(0)	9.8101	9.8101	14.0066	15.4337	16.1555
z(0)	-9.0254	-9.0254	-10.0000	-10.0000	-10.0000
x(0.1)	8.3667	8.3677	11.5549	12.7722	13.3878
z(0.1)	-7.6984	-7.6984	-12.0663	-13.1862	-13.7526
x(0.2)	7.1375	7.1375	9.5540	10.4209	10.8798
z(0.2)	-6.5665	-6.5665	-9.8950	-11.3196	-11.9840
x(0.3)	6.0881	6.0881	7.8916	8.5005	8.8100
z(0.3)	-5.6011	-5.6011	-8.2031	-.92411	-9.8258
x(0.4)	5.1930	5.1930	6.5116	6.9288	7.1292
z(0.4)	-4.7776	-4.776	-6.7949	-7.5523	-7.9696
x(0.5)	4.4295	4.4295	5.6234	5.6435	5.7683
z(0.5)	-4.0752	-4.0752	-5.6234	-6.1676	-6.4510
x(0.6)	3.7782	3.7782	4.4178	4.5931	4.6671
z(0.6)	-3.4760	-3.4760	-4.6495	-5.0330	-5.2199
x(0.7)	3.2227	3.2227	3.6319	3.7354	3.7761
z(0.7)	-2.9649	-2.9649	-3.8404	-4.1041	-4.2234
x(0.8)	2.7489	2.7489	2.9816	3.0355	3.0552
z(0.8)	-2.5290	-2.5290	-3.1687	-3.3440	-3.4171
x(0.9)	2.3447	2.3447	2.4440	2.4649	2.4719
z(0.9)	-2.1572	-2.1572	-2.6116	-2.7226	-2.7648
x(1.0)	2.0000	2.0000	2.0000	2.0000	2.0000
z(1.0)	-1.8400	-1.8400	-2.1497	-2.2149	-2.2369

Note: (i) The degenerate and zeroth order solutions of x(k) are the
same
   (ii) The degenerate and zeroth order solution of z(k) are the
same except at k = 0.
   (iii) The exact solution is obtained by solving the
differential equation.

TABLE - 2.3b

Comparison of approximate and exact solutions of Example 2.3
(Inside the boundary layer)

$x(t)/z(t)$	Degenerate solution	Zeroth order solution	1st order solution	2nd order solution	Exact solution
$x(0)$*	9.8101	9.8101	14.0066	15.4337	16.1555
$z(0)$*	-9.0254	-10.0000	-10.0000	-10.0000	-10.0000
$x(0.01)$	9.6220	9.6132	13.7641	15.1757	15.8895
$z(0.01)$	-9.1500	-9.9471	-10.7036	-10.9609	-11.0910
$x(0.02)$	9.4348	9.4189	13.5178	14.9118	15.6167
$z(0.02)$	9.2180	-9.8683	-11.2354	-11.7003	-11.9354
$x(0.03)$	9.2491	9.2275	13.2695	14.6440	15.3392
$z(0.03)$	-9.2400	-9.7689	-11.6259	-12.2574	-12.5768
$x(0.04)$	9.0652	9.0391	13.0203	14.3742	15.0589
$z(0.04)$	-9.2243	-9.6531	-11.9005	-12.6648	-13.0513
$x(0.05)$	8.8834	8.8537	12.7714	14.1037	14.7775
$z(0.05)$	-9.1784	-9.5244	-12.0802	-12.9493	-13.3889

*The values of $x(0)$ and $z(0)$ for the solution inside the boundary
layer are obtained from Table 2.3a.

Note: The exact solution is obtained by solving the differnetial
equation.

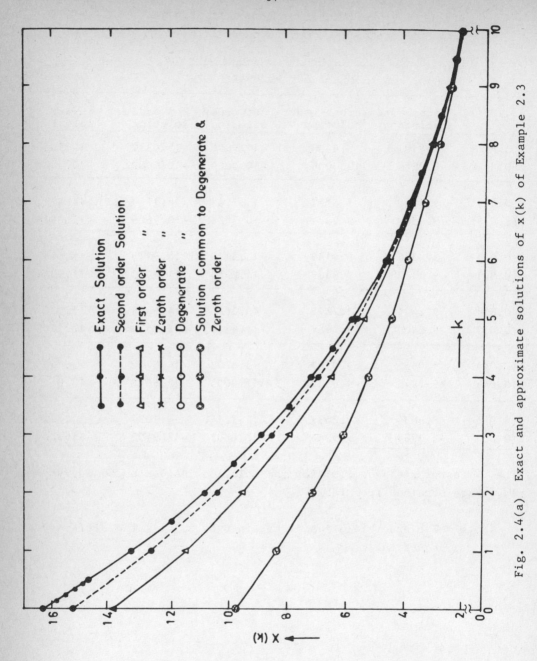

Fig. 2.4(a) Exact and approximate solutions of x(k) of Example 2.3

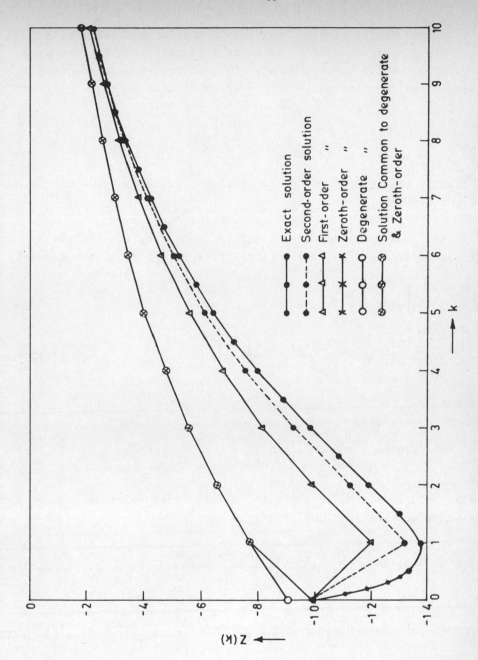

Fig. 2.4(b) Exact and approximate solutions of z(k) of Example 2.3

It is noted that if in the C-model, z(k) is replaced with z(k)/h, the R-model (2.33a) is obtained. Based on this relationship, the whole analysis of the R-model is simplified. To gain proper understanding, let us again consider a second order equation of this model.

$$
\begin{bmatrix} x(k+1) \\ z(k+1) \end{bmatrix} = \begin{bmatrix} a_{11} & a_{12} \\ ha_{21} & ha_{22} \end{bmatrix} \begin{bmatrix} x(k) \\ z(k) \end{bmatrix} \qquad \cdots \quad (2.34)
$$

Evidently, the above model (2.34) has the same eigenvalues as the second order C-model (2.18).

Given x(0) and z(0), the zeroth order solution of the initial value problem (IVP) of (2.34) is obtained from (2.19) as in the C-model

$$
x(k) = a_{11}^{\ k} \left[x(0) + \frac{a_{12}}{a_{11}} z(0) \right] +
$$

$$
h^k \left[- \frac{a_{12}}{a_{11}} z(0) \left\{ \frac{a_{11}a_{22} - a_{12}a_{21}}{a_{11}} \right\}^k \right] \qquad \cdots \quad (2.35a)
$$

$$
z(k) = 0 +
$$

$$
h^k \left[z(0) \left\{ \frac{a_{11}a_{22} - a_{12}a_{21}}{a_{11}} \right\}^k \right] \qquad \cdots \quad (2.35b)
$$

Similarly, given x(N) ad z(0), the zeroth order solution of the boundary value problem (BVP) of (2.34) is

$$
x(k) = a_{11}^{\ k-N} x(N) + h^k \left[- \frac{a_{12}}{a_{11}} \left\{ \frac{a_{11}a_{22} - a_{12}a_{21}}{a_{11}} \right\}^k z(0) \right] \qquad \cdots \quad (2.36a)
$$

$$
z(k) = \qquad 0 \qquad + h^k \left[z(0) \left\{ \frac{a_{11}a_{22} - a_{12}a_{21}}{a_{11}} \right\}^k \right] \qquad \cdots \quad (2.36b)
$$

The degenerate form of (2.34) is

$$
x^{(0)}(k+1) = a_{11} x^{(0)}(k) \qquad \cdots \quad (2.37a)
$$

$$
z^{(0)}(k+1) = \qquad 0 \qquad \cdots \quad (2.37b)
$$

For the IVP, if (2.37a) is solved with $x^{(0)}(0) = x(0) + \dfrac{a_{12}}{a_{11}} z(0)$, the solution of (2.37) is

$$x^{(0)}(k) = \left[x(0) + \frac{a_{12}}{a_{11}} z(0) \right] \left[a_{11} \right]^k \qquad \ldots \quad (2.37c)$$

$$z^{(0)}(k) = 0 \qquad \ldots \quad (2.37d)$$

Note that the solutions $x^{(0)}(k)$ and $z^{(0)}(k)$ obtained above are the same as the terms not involving h^k in (2.35a) and (2.35b) respectively.

For the BVP, if (2.37a) is solved with $x^{(0)}(N) = x(N)$, the solution of (2.37) is

$$x^{(0)}(k) = a_{11}^{k-N} x(N) \qquad \ldots \quad (2.37e)$$

$$z^{(0)}(k) = 0 \qquad \ldots \quad (2.37f)$$

Note again that $x^{(0)}(k)$ and $z^{(0)}(k)$ obtained for BVP of (2.37) are the same terms not involving h^k in (2.36a) and (2.36b) respectively.

A close examination of (2.35) to (2.37) reveals that

(i) As h tends to zero, the uniform convergence of the solutions of $x(k)$ and $z(k)$ to the degenerate solutions $x^{(0)}(k)$ and $z^{(0)}(k)$ respectively fails at $k = 0$, leading to the formation of the boundary layer at $k = 0$.

(II) In the IVP, while the solution of the degenerate equation alone does not satisfy any of the given initial conditions, the combination of degenerate and correction solutions satisfy both the given conditions exactly. In the case of BVP, the degenerate equation satisfies the terminal condition $x(N)$ and drops the initial condition $z(0)$.

(iii) The structures of (2.35) and (2.36) suggest that the transformations required for the boundary layer correction series for both IVP and BVP are

$$v(k) = x(k)/h^k \; ; \; w(k) = z(k)/h^k \qquad\qquad \ldots \; (2.38)$$

It may be noted that while in C-model the transformations for the corrections of $x(k)$ and $z(k)$ are h^{k+1} and h^k, in this model for both $x(k)$ and $z(k)$ the transformations are h^k only.

Using (2.38) in (2.34), the equation of the correction series is obtained as

$$
\begin{bmatrix} hv(k+1) \\ w(k+1) \end{bmatrix}
=
\begin{bmatrix} a_{11} & a_{12} \\ a_{21} & a_{22} \end{bmatrix}
\begin{bmatrix} v(k) \\ w(k) \end{bmatrix}
$$

The degenerate form of the correction series is

$$w^{(0)}(k+1) = \left[\frac{a_{11}a_{22} - a_{12}a_{21}}{a_{11}} \right] w^{(0)}(k) \qquad\qquad \ldots \; (2.39a)$$

$$v^{(0)}(k) = - \frac{a_{12}}{a_{11}} w^{(0)}(k) \qquad\qquad \ldots \; (2.39b)$$

If (2.39) is solved with $w^{(0)}(0) = z(0)$, the solutions are

$$w^{(0)}(k) = \left[\frac{a_{11}a_{22} - a_{12}a_{21}}{a_{11}} \right]^k z(0) \qquad\qquad \ldots \; (2.40a)$$

$$v^{(0)}(k) = - \frac{a_{12}}{a_{11}} \left[\frac{a_{11}a_{22} - a_{12}a_{21}}{a_{11}} \right]^k z(0) \qquad\qquad \ldots \; (2.40b)$$

Note that the solutions $v^{(0)}(k)$ and $w^{(0)}(k)$ obtained above are the same terms associated with h^k in (2.36a) and (2.36b) respectively for the IVP and in (2.37a) and (2.37b) respectively for the BVP.

The total zeroth order series solution is therefore written as

$$x(k) = x^{(0)}(k) + h^k v^{(0)}(k) \qquad\qquad \ldots \; (2.41a)$$

$$z(k) = z^{(0)}(k) + h^k w^{(0)}(k) \qquad\qquad \ldots \; (2.41b)$$

From (2.40) and (2.41), it is clear why $x^{(0)}(0) = x(0) + \dfrac{a_{12}}{a_{11}} z(0)$.

Thus it is evident that though the solution of the degenerate equation does not satisfy any of the given initial conditions, the total series solution (2.4a) satisfies both the given initial conditions exactly.

Analysis of higher order models

Based on the results of the analysis of the second order system, assuming the same transformations h^k for the correction series, the total series solution of (2.33) is taken as

$$x(k) = x_t(k) + h^k v(k) \qquad \qquad \qquad \text{... (2.42a)}$$

$$z(k) = z_t(k) + h^k w(k) \qquad \qquad \qquad \text{... (2.42b)}$$

where $x_t(k)$, $z_t(k)$ and $v(k)$, $w(k)$ refer to the outer and correction series respectively.

Substituting (2.42) in (2.33a) and separating the outer series and the correction series terms, the outer series equation is obtained to be the same as the original equation (2.33a). The correction series equation is

$$\begin{bmatrix} hv(k+1) \\ w(k+1) \end{bmatrix} = \begin{bmatrix} A_{11} & A_{12} \\ A_{21} & A_{22} \end{bmatrix} \begin{bmatrix} v(k) \\ w(k) \end{bmatrix}$$

Note that the correction series equation obtained above is the same as in C-model (2.29).

Assuming series solutions with ascending powers of h, and proceeding as in C-model, the series equations for various orders of approximations are obtained as shown below.

Outer series

$$h^0 : \quad x^{(0)}(k+1) \quad = \quad A_{11}x^{(0)}(k) + E\,u(k)$$

$$z^{(0)}(k+1) \quad = \quad 0$$

$$h^1 : \quad x^{(1)}(k+1) \quad = \quad A_{11}x^{(1)}(k) + A_{12}z^{(1)}(k)$$

$$z^{(1)}(k+1) \quad = \quad A_{21}x^{(0)}(k) + A_{22}z^{(0)}(k) + F\,u(k)$$

$$\cdot\,\cdot\,\cdot\,\cdot \qquad\qquad \cdot\,\cdot\,\cdot\,\cdot\,\cdot\,\cdot\,\cdot\,\cdot\,\cdot\,\cdot\,\cdot\,\cdot\,\cdot$$

$$h^j : \quad x^{(j)}(k+1) \quad = \quad A_{11}x^{(j)}(k) + A_{12}z^{(j)}(k)$$

$$z^{(j)}(k+1) \quad = \quad A_{21}x^{(j-1)}(k) + A_{22}z^{(j-1)}(k) \qquad \ldots \text{ (2.43a)}$$

Correction series

$$h^0 : \qquad 0 \qquad = \quad A_{11}v^{(0)}(k) + A_{12}w^{(0)}(k)$$

$$w^{(0)}(k+1) \quad = \quad A_{21}v^{(0)}(k) + A_{22}w^{(0)}(k)$$

$$h^1 : \quad v^{(0)}(k+1) \quad = \quad A_{11}v^{(1)}(k) + A_{12}w^{(1)}(k)$$

$$w^{(1)}(k+1) \quad = \quad A_{21}v^{(1)}(k) + A_{22}w^{(1)}(k)$$

$$\cdot\,\cdot \quad \cdot\,\cdot\,\cdot\,\cdot\,\cdot \qquad \cdot\,\cdot\,\cdot\,\cdot\,\cdot\,\cdot\,\cdot\,\cdot\,\cdot\,\cdot\,\cdot\,\cdot$$

$$h^j : \quad v^{(j-1)}(k+1) = \quad A_{11}v^{(j)}(k) + A_{12}w^{(j)}(k)$$

$$w^{(j)}(k+1) \quad = \quad A_{21}v^{(j)}(k) + A_{22}w^{(j)}(k)$$

The total series solution is

$$x(k) \quad = \quad \sum_{r=0}^{\infty} \; [x^{(r)}(k) + h^k v^{(r)}(k)]\, h^r \qquad \ldots \text{ (2.44a)}$$

$$z(k) \quad = \quad \sum_{r=0}^{\infty} \; [z^{(r)}(k) + h^k w^{(r)}(k)]\, h^r \qquad \ldots \text{ (2.44b)}$$

The above series equations, total series solution etc. are the same for the initial and boundary value problems. But the selection of the boundary values for the series equation is different and given below.

Boundary conditions for solving the series equations

The boundary conditions required for solving the series equations (2.43) are easily obtained from (2.44).

For the IVP given $x(0)$ and $z(0)$, the initial conditions of the series equations in the sequence in which they are to be solved is given by

$$x^{(0)}(0) = x(0) + A_{11}^{-1} A_{12}(0)$$

$$w^{(0)}(0) = z(0)$$

$$w^{(1)}(0) = -z^{(1)}(0)$$

$$x^{(1)}(0) = -v^{(1)}(0)$$

$$\cdots \qquad \cdots \cdots \cdots$$

$$w^{(j)}(0) = -z^{(j)}(0)$$

$$x^{(j)}(0) = -v^{(j)}(0)$$

For the BVP given x(N) and z(0), the boundary conditions in the sequence in which they have to be solved are

$$x^{(0)}(N) = x(N)$$

$$w^{(0)}(0) = z(0)$$

$$x^{(1)}(N) = 0$$

$$w^{(1)}(0) = -z^{(1)}(0)$$

. . . .

$$x^{(j)}(N) = 0$$

$$w^{(j)}(0) = -z^{(j)}(0)$$

Note: It is important to note that all the outer series equations (2.43a) are of order n, while the correction series equations are of order m. Further, it is essential that A_{11} is non-singular for being able to solve these equations.

Example 2.4: R-model

Consider the fourth order inital value problem [24]

$$\begin{bmatrix} x_1(k+1) \\ x_2(k+1) \\ z_1(k+1) \\ z_2(k+1) \end{bmatrix} = \begin{bmatrix} 1.0 & 0.6 & 0.4 & 0.8 \\ -0.2 & 1.3 & 0.0 & 0.7 \\ 0.0h & 1.0h & 0.5h & 1.0h \\ -1.0h & 1.0h & 0.0h & 1.0h \end{bmatrix} \begin{bmatrix} x_1(k) \\ x_2(k) \\ z_1(k) \\ z_2(k) \end{bmatrix}$$

with the initial conditions

$$\begin{bmatrix} x_1(0) \\ x_2(0) \end{bmatrix} = \begin{bmatrix} 1.0 \\ 0.8 \end{bmatrix} \quad ; \quad \begin{bmatrix} z_1(0) \\ z_2(0) \end{bmatrix} = \begin{bmatrix} 0.6 \\ 0.5 \end{bmatrix}$$

and h = 0.1

The eigenvalues of the above model are

$$P_{1,2} = 1.16 \pm j0.37 \; ; \; p_3 = 0.1 \; ; \; p_4 = 0.0235$$

The eigenvalues of the degenerate model are

$$P_{1,2} = 1.15 \pm j0.31225$$

Using the method developed, the series equations for the degenerate, zeroth, first and second order approximations are solved and the results are compared with the exact solution in Table 2.4. With the increased order of approximation, the series solutions approach the exact solution very closely.

2.7 Analysis of D-model [30,31]

Consider the singularly perturbed (m+n)th order D-model (2.6) reproduced as

$$\begin{bmatrix} x(k+1) \\ hz(k+1) \end{bmatrix} = \begin{bmatrix} A_{11} & A_{12} \\ A_{21} & A_{22} \end{bmatrix} \begin{bmatrix} x(k) \\ z(k) \end{bmatrix} + \begin{bmatrix} E \\ F \end{bmatrix} u(k) \quad \ldots (2.45)$$

Letting h = 0, the degenerate model of (2.45) is obtained as

$$x^{(0)}(k+1) = [\, A_{11} - A_{12}A_{22}^{-1}A_{21}\,]x^{(0)}(k) + [E - A_{12}A_{22}^{-1}F]u(k) \quad \ldots (2.46a)$$

$$z^{(0)}(k) = -A_{22}^{-1}A_{21}x^{(0)}(k) - A_{22}^{-1}F \, u(k) \quad \ldots (2.46b)$$

In the above equations, (2.46a) is a difference equation of order m, while (2.46b) is an algebraic relationship between $x^{(0)}(k)$ and $z^{(0)}(k)$.

104

TABLE - 2.4a

Comparison of approximate and exact solutions of $x_1(k)$ of
Example 2.4

k	Degenerate solution	Zeroth order solution	1st order solution	2nd order solution	Exact solution
0	1.4380	1.0000	1.0000	1.0000	1.0000
1	2.1200	2.1200	2.1200	2.1200	2.1200
2	2.8340	2.8340	2.9220	2.9220	2.9220
3	3.5078	3.5078	3.5816	3.5884	3.5884
4	4.0436	4.0436	3.9922	3.9870	3.9876
5	4.3193	4.3193	3.9705	3.9286	3.9271
6	4.1925	4.1925	3.3106	3.1987	3.1914
7	3.5092	3.5092	1.8035	1.5865	1.5699
8	2.1179	2.1179	-0.7330	-1.0798	-1.1062
9	-0.1119	-0.1119	-4.4161	-4.8864	-4.9152
10	-3.2647	-3.2647	-9.2498	-9.7784	-9.7877
11	-7.3501	-7.3501	-15.0702	-15.4989	-15.4453
12	-12.2692	-12.2692	-21.4906	-21.5302	-21.3440
13	-17.7821	-17.7821	-27.8506	-27.0479	-26.6239
14	-23.4765	-23.4765	-33.1819	-30.8994	-30.1431

Note: The degenerate and zeroth order solutions are the same except at k = 0. The first and second order solutions coincide with the exact solution upto k = 2 and k = 3 respectively. This is for the similar reason given under note in Example 1.1. Similar behaviour is seen for the other state variables also.

TABLE - 2.4b

Comparison of approximate and exact solutions of $x_2(k)$
of Example 2.4

k	Degenerate solution	Zeroth order solution	1st order solution	2nd order solution	Exact solution
0	1.1366	0.8000	0.8000	0.8000	0.8000
1	1.1900	1.1900	1.1900	1.1900	1.1900
2	1.1230	1.1230	1.1440	1.1440	1.1440
3	0.8931	0.8931	0.8377	0.8398	0.8398
4	0.4595	0.4595	0.2529	0.2431	0.2433
5	-0.2114	-0.2114	-0.6527	-0.6854	-0.6867
6	-1.1387	-1.1387	-1.8935	-1.9568	-1.9608
7	-2.3188	-2.3188	-3.4408	-3.5323	-3.5385
8	-3.7163	-3.7163	-5.2070	-5.3054	-5.3099
9	-5.2548	-5.2548	-7.0304	-7.0854	-7.0787
10	-6.8089	-6.8089	-8.6647	-8.5877	-8.5533
11	-8.1986	-8.1986	-9.7741	-9.4322	-9.3466
12	-9.1881	-9.1881	-9.9404	-9.1571	-8.9935
13	-9.4907	-9.4907	-8.6838	-7.2523	-6.9891
14	-8.7815	-8.7815	-5.5031	-3.2159	-2.8513

Note: The degenerate and zeroth order solutions are the same except at k = 0.

TABLE - 2.4c

Comparison of approximate and exact solution of $z_1(k)$ of
Example 2.4

k	Degenerate solution	Zeroth order solution	1st order solution	2nd order solution	Exact solution
0	0.0000	0.6000	0.6000	0.6000	0.6000
1	0.0000	0.0000	0.1600	0.1600	0.1600
2	0.0000	0.0000	0.1190	0.1300	0.1300
3	0.0000	0.0000	0.1123	0.1110	0.1119
4	0.0000	0.0000	0.0893	0.0723	0.0709
5	0.0000	0.0000	0.0459	0.0036	-0.0015
6	0.0000	0.0000	-0.0211	-0.0988	-0.1091
7	0.0000	0.0000	-0.1139	-0.2357	-0.2417
8	0.0000	0.0000	-0.2319	-0.4031	-0.4230
9	0.0000	0.0000	-0.3716	-0.5906	-0.6089
10	0.0000	0.0000	-0.5255	-0.7800	-0.7860
11	0.0000	0.0000	-0.6809	-0.9442	-0.9210
12	0.0000	0.0000	-0.8199	-1.0469	-0.9710
13	0.0000	0.0000	-0.9188	-1.0435	-0.8859
14	0.0000	0.0000	-0.9491	-0.8835	-0.6135

Note: The degenerate and zeroth order solutions are the same except at k = 0.

TABLE 2.4d

Comparison of approximate and exact solutions of $z_2(k)$ of
Example 2.4

k	Degenerate solution	Zeroth order solution	1st order solution	2nd order solution	Exact solution
0	0.0000	0.5000	0.5000	0.5000	0.5000
1	0.0000	0.0000	0.0300	0.0300	0.0300
2	0.0000	0.0000	-0.0930	-0.0900	-0.0900
3	0.0000	0.0000	-0.1711	-0.1871	-0.1868
4	0.0000	0.0000	-0.2615	-0.2915	-0.2935
5	0.0000	0.0000	-0.3584	-0.4001	-0.4038
6	0.0000	0.0000	-0.4531	-0.4981	-0.5018
7	0.0000	0.0000	-0.5331	-0.5657	-0.5654
8	0.0000	0.0000	-0.5828	-0.5777	-0.5674
9	0.0000	0.0000	-0.5834	-0.5057	-0.4771
10	0.0000	0.0000	-0.5143	-0.3187	-0.2641
11	0.0000	0.0000	-0.3544	0.0071	0.0970
12	0.0000	0.0000	-0.0849	0.4942	0.6196
13	0.0000	0.0000	0.3081	1.1465	1.2970
14	0.0000	0.0000	0.8291	1.9475	2.0941

Note: The degenerate and zeroth order solutions are the same except at k = 0.

For the initial value problem, given x(0) and z(0), it is easy to infer that the solution is obtained as a power series with terms h^0, h^{-1}, h^{-2},....etc. and the solution of (2.45) is always unstable. Therefore for this model, the singular perturbation method cannot be developed for the IVP. However, for the final and boundary vaue problems, it is possible to develop the method, as shown below.

Consider the second order homogeneous equation

$$
\begin{bmatrix} x(k+1) \\ hz(k+1) \end{bmatrix} = \begin{bmatrix} a_{11} & a_{12} \\ a_{21} & a_{22} \end{bmatrix} \begin{bmatrix} x(k) \\ z(k) \end{bmatrix} \qquad \ldots \quad (2.47)
$$

Given x(N) and z(N) as the terminal or final vaues of x(k) and z(k) respectively, the solution of the final value problem of (2.47) is

$$
\begin{bmatrix} x(k) \\ z(k) \end{bmatrix} = \begin{bmatrix} a_{11} & a_{12} \\ \dfrac{a_{21}}{h} & \dfrac{a_{22}}{h} \end{bmatrix}^{k-N} \begin{bmatrix} x(N) \\ z(N) \end{bmatrix}
$$

$$
= \begin{bmatrix} A \end{bmatrix}^{k-N} \begin{bmatrix} x(N) \\ z(N) \end{bmatrix} \qquad \ldots \quad (2.48a)
$$

Using the Sylvester's formula [13].

$$
A^{k-N} = \frac{p_1^{\,k-N}[\,A - p_2 I\,] - p_2^{\,k-N}[\,A - p_1 I\,]}{p_1 - p_2} \qquad \ldots \quad (2.48b)
$$

where p_1 and p_2 are the eigenvalues of A given by

$$
p_{1,2} = \frac{(a_{11}+a_{22}/h)}{2} \pm \frac{(a_{11}+a_{22}/h)}{2} \left[1 - \frac{4h(a_{11}a_{22}-a_{12}a_{21})}{(a_{11}h+a_{22})^2} \right]^{0.5} \qquad \ldots \quad (2.49)
$$

Using (2.48b) and (2.49) in (2.48a), the zeroth order solution (ignoring terms with h and higher powers of h) of (2.47) is

$$x(k) = \left[\frac{a_{11}a_{22}-a_{12}a_{21}}{a_{11}}\right]^{k-N} x(N) +$$

$$h^{N-k+1}\left[\frac{a_{12}}{a_{22}}(\frac{a_{21}}{a_{22}} x(N) + z(N))(a_{22})^{k-N}\right] \qquad \ldots \quad (2.50a)$$

$$z(k) = -\frac{a_{21}}{a_{22}}\left[\frac{a_{11}a_{22}-a_{12}a_{21}}{a_{22}}\right]^{k-N} x(N) +$$

$$h^{N-k}\left[\{z(N) + \frac{a_{21}}{a_{22}} x(N)\}\{a_{22}\}^{k-N}\right] \qquad \ldots \quad (2.50b)$$

Given $x(0)$ and $z(N)$, for the BVP, the zeroth order solution of (2.47) is obtained from (2.50) as shown below.

Letting $k = 0$ in (2.50a),

$$x(N) = \left[\frac{a_{11}a_{22}- a_{12}a_{21}}{a_{22}}\right]^{N} x(0) + 0(h)$$

Substituting the above value of $x(N)$ in (2.50), the solution of the BVP of (2.47) is

$$x(k) = \left[\frac{a_{11}a_{22} - a_{12}a_{21}}{a_{22}}\right]^{k} x(0) +$$

$$h^{N-k+1}\left[\frac{a_{12}}{a_{22}}\left\{\frac{a_{21}}{a_{22}}(\frac{a_{11}a_{22} - a_{12}a_{21}}{a_{22}})^{N} x(0)(+z(N)\right\}\{a_{22}\}^{k-N}\right]$$
$$\ldots \quad (2.51b)$$

$$z(k) = -\frac{a_{21}}{a_{22}}\frac{a_{11}a_{22}- a_{12}a_{21}}{a_{22}}^{k} x(0) +$$

$$h^{N-k}\left[\{z(N) + \frac{a_{21}}{a_{22}}(\frac{a_{11}a_{22} - a_{12}a_{21}}{a_{22}})^{N} x(0)\}\{a_{22}\}^{k-N}\right]\ldots \quad (2.51b)$$

108

The degenerate form of (2.47) is

$$x^{(0)}(k+1) = \left[\frac{a_{11}a_{22} - a_{12}a_{21}}{a_{22}} \right] x^{(0)}(k) \qquad \ldots \quad (2.52a)$$

$$z^{(0)}(k) = - \frac{a_{21}}{a_{22}} x^{(0)}(k) \qquad \ldots \quad (2.52b)$$

A close examination of (2.50) to (2.52) reveals that

(i) If the degenerate equation (2.52) is solved with $x^{(0)}(N) = x(N)$ for the final value problem, and with $x^{(0)}(0) = x(0)$ for the BVP, the resulting solutions are the same as the first terms (not involving h^{N-k+1} or h^{N-k}) in (2.50) and (2.51) respectively.

(ii) As h tends to zero, the solution $z(k)$ fails to uniformly converge to the degenerate solution $z^{(0)}(k)$ at k=N. This leads to the formation of the boundary layer at k = N with the consequent loss of the boundary condition z(N) in the process of degeneration.

(iii) The structures of (2.50) and (2.51) reveal that the transformations required for the correction series are

$$v(k) = x(k)/h^{N-k+1} \; ; \quad w(k) = z(k)/h^{N-k} \qquad \ldots \quad (2.53)$$

Using the above transformation in (2.47), the equation of the correction series is

$$\begin{bmatrix} v(k+1) \\ w(k+1) \end{bmatrix} = \begin{bmatrix} ha_{11} & a_{12} \\ ha_{21} & a_{22} \end{bmatrix} \begin{bmatrix} v(k) \\ w(k) \end{bmatrix}$$

If the degenerate form of the above correction series equation is solved with $w^{(0)}(N) = z(N) - z^{(0)}(N)$ for the final as well as boundary value problems, the resulting solution $v^{(0)}(k)$ and $w^{(0)}(k)$ are found to be the same as those neglected in the degeneration process (involving h^{N-k+1} or h^{N-k}) in (2.50) and (2.51) respectively. This illustrates the recovery of the lost boundary condition.

Analysis of higher order models

Based on the results of the analysis of second order system and taking the same transformations (2.53) for the correction series, the total series solution of (2.5) is written as

$$x(k) = x_t(k) + h^{N-k+1} v(k) \qquad \ldots \quad (2.54a)$$

$$z(k) = z_t(k) + h^{N-k} w(k) \qquad \ldots \quad (2.54b)$$

where $x_t(k)$, $z_t(k)$ refer to the outer series and $v(k)$, $w(k)$ refer to the correction series.

Substituting (2.54) in (2.55) and separating the outer series and the correction series terms, the outer series equation is obtained as

$$\begin{bmatrix} v(k+1) \\ w(k+1) \end{bmatrix} = \begin{bmatrix} hA_{11} & A_{12} \\ hA_{21} & A_{22} \end{bmatrix} \begin{bmatrix} v(k) \\ w(k) \end{bmatrix} \qquad \ldots \quad (2.55)$$

Assuming series solutions with ascending powers of h for both (2.54) and (2.55) and proceeding in the usual manner, the series equations for the outer and correction series are obtained as shown below.

Outer series

$$h^0 : \quad x^{(0)}(k+1) \quad = A_{11}x^{(0)}(k) + A_{12}z^{(0)}(k) + E\,u(k)$$

$$0 \quad = A_{21}x^{(0)}(k) + A_{22}z^{(0)}(k) + F\,u(k)$$

$$h^1 : \quad x^{(1)}(k+1) \quad = A_{11}x^{(1)}(k) + A_{12}z^{(1)}(k)$$

$$z^{(0)}(k+1) \quad = A_{21}x^{(1)}(x) = A_{22}z^{(1)}(k)$$

$$\cdots \qquad \cdots$$

$$h^j : \quad x^{(j)}(k+1) \quad = A_{11}x^{(j)}(k) + A_{12}z^{(j)}(k)$$

$$z^{(j-1)}(k+1) \quad = A_{21}x^{(j)}(k) + A_{22}z^{(j)}(k) \qquad \cdots \quad (2.56a)$$

Correction series

$$h^0 : \quad v^{(0)}(k+1) \quad = A_{12}w^{(0)}(k)$$

$$w^{(0)}(k+1) \quad = A_{22}w^{(0)}(k)$$

$$h^1 : \quad v^{(1)}(k+1) \quad = A_{11}v^{(0)}(k) + A_{12}w^{(1)}(k)$$

$$w^{(1)}(k+1) \quad = A_{21}v^{(0)}(k) + A_{22}w^{(1)}(k)$$

$$\cdots \qquad \cdots$$

$$h^j : \quad v^{(j)}(k+1) \quad = A_{11}v^{(j-1)}(k) + A_{12}w^{(j)}(k)$$

$$w^{(j)}(k+1) \quad = A_{21}v^{(j-1)}(k) + A_{22}w^{(j)}(k) \qquad \cdots \quad (2.56b)$$

Boundary conditions of series equations

The total series solution of (2.45) is

$$x(k) = \sum_{r=0}^{\infty} [x^{(r)}(k) + h^{N-k+1}v^{(r)}(k)]\, h^r \qquad \cdots \quad (2.57a)$$

$$z(k) = \sum_{r=0}^{\infty} [z^{(r)}(k) + h^{N-k}w^{(r)}(k)] h^r \qquad \ldots \quad (2.57b)$$

The boundary conditions required to solve the series equations (2.56) are readily obtained from (2.57) as shown below.

For the <u>final value problem</u>, given x(N) and z(N)

$$x^{(0)}(N) = x(N) \qquad ; \quad w^{(0)}(N) = z(N) - z^{(0)}(N)$$

$$x^{(1)}(N) = -v^{(0)}(N) \qquad ; \quad w^{(1)}(N) = -z^{(1)}(N)$$

. . . .

$$x^{(j)}(N) = -v^{(j-1)}(N) \quad ; \quad w^{(j)}(N) = -z^{(j)}(N)$$

For the <u>boundary value problem</u>, given x(0) and z(N),

$$x^{(0)}(0) = x(0) \qquad ; \quad w^{(0)}(N) = z(N) - z^{(0)}(N)$$

$$x^{(1)}(0) = 0 \qquad ; \quad w^{(1)}(N) = - z^{(1)}(N)$$

. . . .

$$x^{(j)}(0) = 0 \qquad ; \quad w^{(j)}(N) = - z^{(j)}(N)$$

<u>Note</u>: It is seen that all the above outer series equations (2.56a) are of order m, while the correction series equations (2.56b) are of order n. It is essential that A_{22} is non-singular for these equation to be solved. For the correction series it is enough if the solutions are found for a few values of k near the boundary layter (k = N) due to the transformations h^{N-k+1} and h^{N-k}.

<u>Interrelationship between C- and D-models</u>

The interrelationship between C- and D-models is shown as below. Consider the homogeneous form of the C-model (2.16) reproduced as below.

$$\begin{bmatrix} x(k+1) \\ z(k+1) \end{bmatrix} = \begin{bmatrix} A_{11} & hA_{12} \\ A_{21} & hA_{22} \end{bmatrix} \begin{bmatrix} x(k) \\ z(k) \end{bmatrix}$$

or

$$\begin{bmatrix} x(k+1) \\ \\ z(k+1) \end{bmatrix} = \begin{bmatrix} A_{11} & hA_{12} \\ \\ A_{21} & hA_{22} \end{bmatrix} \begin{bmatrix} x(k) \\ \\ z(k) \end{bmatrix}$$

The above matrix equation can be written as

$$\begin{bmatrix} x(k) \\ \\ hz(k) \end{bmatrix} = \begin{bmatrix} \bar{A}_{11} & \bar{A}_{12} \\ \\ \bar{A}_{21} & \bar{A}_{22} \end{bmatrix} \begin{bmatrix} x(k+1) \\ \\ z(k+1) \end{bmatrix}$$

where

$$\begin{bmatrix} \bar{A}_{11} & \bar{A}_{12} \\ \\ \bar{A}_{21} & \bar{A}_{22} \end{bmatrix} = \begin{bmatrix} A_{11} & A_{12} \\ \\ A_{21} & A_{22} \end{bmatrix}^{-1}$$

Replacing k with N-k, the above equation can be written as

$$\begin{bmatrix} x(N-k+1) \\ \\ hz(N-k+1) \end{bmatrix} = \begin{bmatrix} \bar{A}_{11} & \bar{A}_{12} \\ \\ \bar{A}_{21} & \bar{A}_{22} \end{bmatrix} \begin{bmatrix} x(N-k) \\ \\ z(N-k) \end{bmatrix}$$

which is evidently in D-model.

Proceeding on similar lines, it can be shown that the C-model is obtained from D-model

Example 2.5: D-model

Consider the third order two point boundary value problem

$$\begin{bmatrix} x(k+1) \\ \\ hz_1(k+1) \\ \\ hz_2(k+1) \end{bmatrix} = \begin{bmatrix} 0 & 1 & 0 \\ \\ 0 & 0 & 1 \\ \\ 0.6 & -0.89 & 1.83 \end{bmatrix} \begin{bmatrix} x(k) \\ \\ z_1(k) \\ \\ z_2(k) \end{bmatrix} + \begin{bmatrix} 0 \\ \\ 0 \\ \\ 0.1 \end{bmatrix} u(k)$$

with the boundary conditions

$$
x(0) \;=\; 2.5 \quad ; \quad \begin{bmatrix} z_1(8) \\ z_2(8) \end{bmatrix} = \begin{bmatrix} 8.0 \\ 1.0 \end{bmatrix}
$$

h = 0.1 and u(k) is a unit step function.

The eigenvalues of the above model are

$$ p_1 = 10.0 \; ; \; p_2 = 7.5 \; ; \; p_3 = 0.8 $$

The eigenvalue of the degenerate model is

$$ p = 0.6742 $$

Using the method developed, the series equations upto second order approximation are solved and the solutions are plotted in Fig. 2.5, which clearly illustrate that the approximate solutions approach the exact solution with the increased order of approximation.
Note: The above example is the state space version of example 1.4, and the above model is obtained following the method given in Sec. 2.1

2.8 Two-Time-Scale Property

An examination of the eigenvalues and solutions of the three state space models discussed in Sections 2.5 to 2.7 reveals a very important property of singularly perturbed difference equations, that is, the "two-time-scale" property.

(a) C- and R-models

It is apparent from the C-and R-models (2.16) and (2.33a) that they have the same eigenvalues. Hence they exhibit similar two-time-scale properties. For the second order models, for sufficiently small values of h their eigenvalues can be written from (2.20) as

$$ p_1 = a_{11} + \frac{h a_{12} a_{21}}{a_{11}} + 0\,(h^2) \qquad \qquad \ldots \; (2.58a) $$

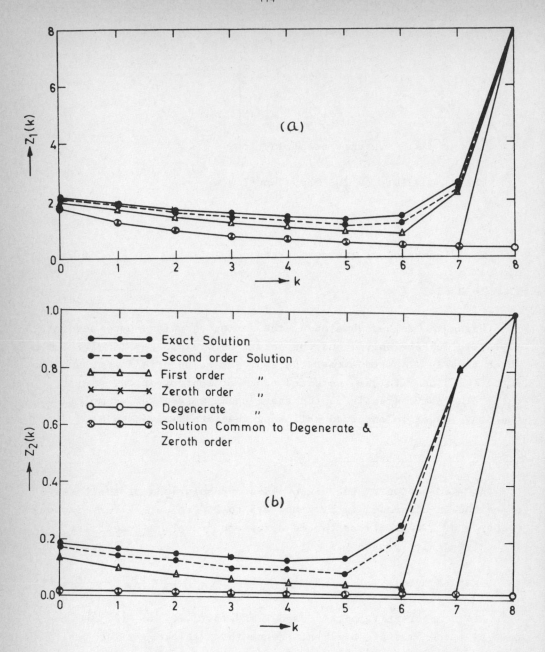

Fig. 2.5(a) & (b)   Exact and approximate solutions of $z_1(k)$ & $z_2(k)$ of
Example 2.5

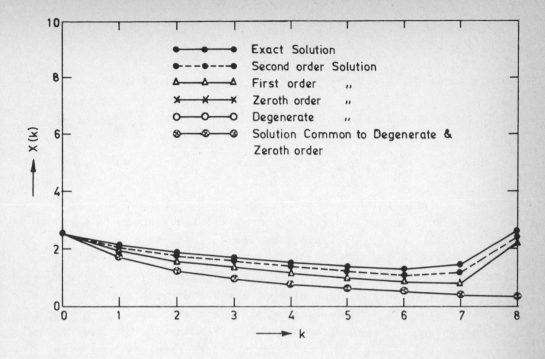

Fig. 2.5(c)   Exact and approximate solutions of x(k) of Example 2.5

$$p_2 = h \; (\frac{a_{11}{}^1{}_{22} - a_{12}a_{21}}{a_{11}}) + 0 \; (h^2) \qquad \qquad \text{... (2.58b)}$$

The solutions of (2.18) and (2.34) can be written as

$$y(k) = y_1(k) + y_2(k) \qquad \qquad \text{... (2.59)}$$

where $y_1(k) = c_1(p_1)^k$ and $y_2(k) = c_2(p_2)^k$

From (2.58), it follows that $|p_1| \gg |p_2|$. Therefore in (2.59), $y_2(k)$ has a much faster variation than $y_1(k)$ with respect to k and we say that (2.18) and (2.34) exhibit the two-time-scale property with $y_1(k)$ and $y_2(k)$ as the slow and fast mode respectively. In the degeneration process, the eigenvalue p_1 associated with the slow mode approaches a_{11} while the eigenvalue p_2 associated with fast mode approaches zero. Thus the fast mode is always stable and contributes to a fast decay of the solution near k = 0 leading to the formation of the boundary layer at k = 0. This is in contrast to the continuous system models where the eigenvalue corresponding to the fast mode approaches minus infinity in the degeneration process.

We can extend these ideas to the general (n+m)th order models (2.16) and (2.33a) as follows. The models possess two distinct groups of eigenvalues widely separated from each other. The slow modes correspond to n eigenvalues which approach the eigenvalues of A_{11} in the process of degeneration. The m eigenvalues corresponding to the fast modes are much smaller in magnitude than the rest and approach zero in the degeneration process. in the complex plane, the n eigenvalues corresponding to the slow modes form a group and lie in the neighbourhood of the circumference of the unit circle. The m eigenvalues corresponding to the fast modes form the group and lie near the origin of the unit circle.

(b) D-Model

For the second order system of D-Model (2.47), the eigenvalues are given by (2.49). For sufficiently small values of h, they are written as

$$p_1 = \frac{a_{11}a_{22} - a_{12}a_{21}}{a_{22}} + 0(h) \qquad \qquad \text{... (2.59a)}$$

$$p_2 = \frac{a_{22}}{h} + \frac{a_{12}a_{21}}{a_{22}} + 0(h) \qquad \qquad \cdots \quad (2.59b)$$

The solution of (2.47) can be written as

$$y(k) = y_1(k) + y_2(k) \qquad \qquad \cdots \quad (2.60)$$

where $y_1(k) = c_1(p_1)^k$ and $y_2(k) = c_2(p_2)^k$

From (2.59), $|p_1| \gg |p_2|$ and it follows from (2.60) that $y_2(k)$ has a much faster variation than $y_1(k)$ with respect to k. Therefore for this model $y_1(k)$ and $y_2(k)$ are called the slow and fast modes respectively. it is noted from (2.59) that in the degeneration process while p_1 approaches ($\frac{a_{11}a_{22} - a_{12}a_{21}}{a_{22}}$), p_2 approaches <u>infinity</u>. Thus the contribution to the solution due to the fast mode corresponds to a fast <u>rise</u> near k = N (terminal point) leading to the formation of the boundary layer at k = N.

For the (m+n)th order model (2.45), it follows that it possesses two distinct groups of eigenvalues with a wide separation between them. The slow modes correspond to m eigenvalues which in the process of degeneration approach the eigenvalues of the matrix $(A_{11} - A_{12} A_{22}^{-1} A_{21})$. The fast modes correspond to n eigenvalues that are much larger in magnitude than those corresponding to the slow modes and approach <u>infinity</u> in the process of degeneration. In the complex plane, eigenvalues corresponding to the slow modes are located near the <u>curcumference</u> of the unit circle and the eigenvalue corresponding to the fast modes are located far <u>outside the circumference</u> of the unit circle.

A comparison of the two-time-scale properties of the three state space models is interesting. In the C-and R-models, the moduli of the eigenvalues of the fast modes are much <u>less than unity</u> and correspond to a fast <u>decay</u> of the solution near k = 0 leading to the formation of the boundary layer at k = 0. On the other hand, in D-model, the moduli of the eigenvalues of the fast modes are much <u>larger than unity</u> and contribute to a fast <u>rise</u> of the solution at k = N, leading to the formation boundary layer at k = N. In all the three models, the moduli of the eigenvalues of the slow modes are in some neighbourhood of one. The slow modes can therefore be either stable or unstable.

2.9 Asymptotic Correctness of Series Expansions [32]

Consider the homogeneous (n+m)th order R-model (2.33a). In order to prove the asymptotic correctness of the "formal series" expansions obtained as in (2.44), it needs to be shown that [8],

$$x_E(k) - \sum_{r=0}^{j} [x^{(r)}(k) + h^k v^{(r)}(k)] h^r = 0(h^{j+1})$$

$$z_E(k) - \sum_{r=0}^{j} [z^{(r)}(k) + h^k w^{(r)}(k)] h^r = 0(h^{j+1})$$

for $j = 0,1,2,......$

where

$x_E(k)$ and $z_E(k)$ are the exact solutions;

$x^{(r)}(k)$ and $z^{(r)}(k)$ are the solutions of outer series;

$v^{(r)}(k)$ and $w^{(r)}(k)$ are the solutions of the correction series of x(k) and z(k) respectively.

j is the order of approximation desired.

The detailed proof with illustration is given in the paper published by the authors [32]. Due to the interrelationships existing between C-, R- and D-models, this proof for the asymptotic correctness is adequate for the other models also.

2.10 Conclusions

In this chapter, investigations are carried out on [29-32]

(a) The formulation of the singularly perturbed difference and differential equations into the three state space discrete-models, i.e. C-model, R-model, and D-model.

(b) The analysis of the above three state space models.

In modelling:

(i) It has been shown how the classical difference equations can be put in the three state space models, and the interrelationship between the models has been given.

(ii) While obtaining the discrete models of differential equations, it has been found that by choosing the discretizing interval slightly higher than the singular perturbation parameter, the discrete models are automatically cast in C- or D-models.

Sampled-data systems have been dealt on similar lines.

In analysis:

(i) For all the models, singular perturbation methods have been developed bringing out the various salient features.

(ii) In examining the two-time-scale property of discrete models, it has been found that C- and R-models are always associated with stable fast modes leading to the formation of the boundary layer at the initial point ($k = 0$). In the D-model the fast modes are unstable and the boundary layer is formed at the terminal point ($k = N$). For all the models, the slow modes can be either stable or unstable.

(iii) Finally, the asymptotic correctness of the series expansions has been established [32].

CHAPTER - 3

THREE-TIME-SCALE DIFFERENCE EQUATIONS WITH APPLICATION

TO OPEN-LOOP OPTIMAL CONTROL PROBLEM

A state space model with three-time-scale property exhibiting boundary layer behaviour at the initial and final points is formulated. The solution of the model is obtained as the sum of an outer series solution and two correction series solutions called the initial and final boundary layer correction series solutions. The optimal control problem with a quadratic cost function is then considered. Using the discrete maximum principle, the state and costate equations are obtained and cast in the singularly perturbed form which exhibits the three-time-scale property. A method is developed to solve the resulting two-point boundary value problem. Examples are provided to illustrate the proposed method [33].

3.1 Three-Time-Scale Difference Equations [30]

In Sec. 2.8, the two-time-scale property of C- and D-models is examined in detail. It is shown that in C-model (2.3) where the small parameter h is associated with $z(k)$, the fast modes have eigenvalues with moduli much less than unity which approach <u>zero</u> in the degeneration process. Consequently, the effect of the fast modes to the total solution is a <u>fast decay</u> at k = 0, leading to the formation of the boundary layer at the initial point (k=0). In D-model (2.6) where the small parameter h is associated with <u>z(k+1)</u>, the fast modes have the eigenvalues with moduli much <u>larger than unity</u> which approach <u>infinity</u> in the degeneration process. Therefore their contribution to the total solution is a <u>fast rise</u> near k = N (terminal or final point), leading to the formation of the boundary layer at the final point (k = N). Keeping these points in view, a state space model which inherits <u>both</u> these features is formulated as below.

$$
\begin{bmatrix} x(k+1) \\ y(k+1) \\ hz(k+1) \end{bmatrix} = \begin{bmatrix} A_{11} & hA_{12} & A_{13} \\ A_{21} & hA_{22} & A_{23} \\ A_{31} & hA_{32} & A_{33} \end{bmatrix} \begin{bmatrix} x(k) \\ y(k) \\ z(k) \end{bmatrix} + \begin{bmatrix} E \\ F \\ G \end{bmatrix} u(k) \quad \dots \quad (3.1)
$$

where

$x(k)$ is (n_1 x 1) state vector;

$y(k)$ is (n_2 x 1) " " ;

$z(k)$ is (n_3 x 1) " " ;

$u(k)$ is (r x 1) " " ;

A_{ij}, i, j = 1,2,3, E, F and G are matrices of compatible dimensionality, and with the boundary conditions

$x(0)$ or $x(N)$, $y(0)$ and $z(N)$.

The degenerate form of (3.1) is

$$x^{(0)}(k+1) = A_{11} x^{(0)}(k) + A_{13} z^{(0)}(k) + E u(k)$$

$$y^{(0)}(k+1) = A_{21} x^{(0)}(k) + A_{23} z^{(0)}(k) + F u(k)$$

$$0 \quad = A_{31} x^{(0)}(k) + A_{33} z^{(0)}(k) + G u(k)$$

The above equation are rearranged as

$$x^{(0)}(k+1) = [A_{11}-A_{13}A_{33}^{-1}A_{31}] x^{(0)}(k) + [E-A_{13}A_{33}^{-1}G] u(k) \quad \ldots \quad (3.2a)$$

$$y^{(0)}(k+1) = [A_{21}-A_{23}A_{33}^{-1}A_{31}] x^{(0)}(k) + [F-A_{23}A_{33}^{-1}G] u(k) \quad \ldots \quad (3.2b)$$

$$z^{(0)}(k) \quad = -A_{33}^{-1}A_{31}x^{(0)}(k) - A_{33}^{-1}G u(k) \quad \ldots \quad (3.2c)$$

In the above three equations only (3.2a) is a difference equation of order n_1 while the other two (3.2b,c) are algebraic relationships. In other words, once $x^{(0)}(k)$ is solved from (3.2a), the solutions $y^{(0)}(k)$ and $z^{(0)}(k)$ are automatically fixed from (3.2b,c). Since the order of the equation of (3.1) drops from $(n_1 + n_2 + n_3)$ to n_1, it is in the singularly perturbed form. It follows from (3.2) that A_{33} should be non-singular to solve (3.2a).

Consider a second order equation with the features of (3.1) as

$$
\begin{bmatrix} y(k+1) \\ hz(k+1) \end{bmatrix} = \begin{bmatrix} a_{11} & a_{12} \\ a_{21} & a_{22} \end{bmatrix} \begin{bmatrix} h\ y(k) \\ z(k) \end{bmatrix}
$$

or

$$
\begin{bmatrix} y(k+1) \\ z(k+1) \end{bmatrix} = \begin{bmatrix} ha_{11} & a_{12} \\ a_{21} & \dfrac{a_{22}}{h} \end{bmatrix} \begin{bmatrix} y(k) \\ z(k) \end{bmatrix} \qquad \dots \ (3.3)
$$

The two eigenvalues of (3.3) are

$$
P_1 = \frac{a_{22}}{h} \ ; \quad P_2 = h\ (\frac{a_{11}\ a_{22} - a_{12}\ a_{21}}{a_{22}})
$$

Given $y(0)$ and $z(N)$, the zeroth order solution of (3.3) is obtained using Sylvester's expansion [13] and proceeding as in the analysis of D-model (2.47). Then we have

$$
y(k) = h^{N-k+1} \underbrace{\begin{bmatrix} a_{12}(a_{22})^{k-N-1}z(N) \end{bmatrix}}_{y_f^{(0)}(k)} + h^k \underbrace{\begin{bmatrix} y(0)(\frac{a_{11}a_{22}-a_{12}a_{21}}{a_{22}})^k \end{bmatrix}}_{y_i^{(0)}(k)} \quad \dots \ (3.4a)
$$

$$
z(k) = h^{N-k} \underbrace{\begin{bmatrix} (a_{22})^{k-N}z(N) \end{bmatrix}}_{z_f^{(0)}(k)} + h^{k+1} \underbrace{\begin{bmatrix} -\frac{a_{21}}{a_{22}}y(0)(\frac{a_{11}a_{22}-a_{12}a_{21}}{a_{22}})^k \end{bmatrix}}_{z_i^{(0)}(k)} \quad \dots \ (3.4b)
$$

Note that the degenerate form of (3.3) is of zero order and hence has a trivial solution. A close examination of (3.4a) shows that as h tends to zero, the uniform convergence of $y(k)$ to the trivial degenerate solution fails at $k = 0$, leading to the formation of the boundary layer at the initial point $(k = 0)$. Similarly, from (3.4b) it follows that as h tends to zero, the uniform convergence of $z(k)$ to the trivial degenerate solution fails at $k = N$, leading to the formation of boundary layer at the final point $(k = N)$. Thus in the process of degeneration, both the boundary conditions $y(0)$ and $z(N)$ are lost.

The structure of (3.4) suggests that the two boundary conditions lost can be recovered by incorporating two separate correction series called the initial boundary layer correction (BLC) series and the final boundary layer correction (BLC) series using the following transformations.

For the initial BLC,

$$y_i(k) = y(k)/h^k \; ; \; z_i(k) = z(k)/h^{k+1} \qquad \ldots \quad (3.5a)$$

For the final BLC,

$$y_f(k) = y(k)/h^{N-k+1} \; ; \; z_f(k) = z(k)/h^{N-k} \qquad \ldots \quad (3.5b)$$

Using (3.5a), the equation for inital BLC is

$$\begin{bmatrix} y_i(k+1) \\ \\ h^2 z_i(k+1) \end{bmatrix} = \begin{bmatrix} a_{11} & a_{12} \\ \\ a_{21} & a_{22} \end{bmatrix} \begin{bmatrix} y_i(k) \\ \\ z_i(k) \end{bmatrix} \qquad \ldots \quad (3.6)$$

Using (3.5b), the equation for final BLC is

$$\begin{bmatrix} y_f(k+1) \\ \\ z_f(k+1) \end{bmatrix} = \begin{bmatrix} h^2 a_{11} & a_{12} \\ \\ h^2 a_{21} & a_{22} \end{bmatrix} \begin{bmatrix} y_f(k) \\ \\ z_f(k) \end{bmatrix} \qquad \ldots \quad (3.7)$$

Taking $y_i^{(0)}(0) = y(0)$, if the degenerate equation of (3.6) is solved, the solutions $y_i^{(0)}(k)$ and $z_i^{(0)}(k)$ obtained are the same as indicated in (3.4). Similarly, taking $z_f^{(0)}(N) = z(N)$, if the degenerate form of (3.7) is solved, the resulting solutions $y_f^{(0)}(k)$ and $z_f^{(0)}(k)$ are the same as shown in (3.4). This clearly demonstrates the recovery of the lost boundary conditions.

Analysis of higher order model

Consider the $(n_1+n_2+n_3)$ order model (3.1). The degenerate form of this model (3.2) is shown to be of order n_1. Utilizing the results of the second order model (3.3), it is apparent that since h is associated with y(k), a boundary layer occurs at k = 0 accounting for the loss of n_2 initial conditions y(0). similarly, since h is associated with z(k+1), another boundary layer occurs at k = N accounting for the loss of n_3 terminal conditions z(N). The degenerate model of (3.2) which is of order n_1, satisfies the boundary conditions fo x(k) only. In order to recover these lost boundary conditions, we incorporate two separate correction series, known as initial BLC and final BLC using the transformations (3.5).

The total series solution is written as

$$
\left.
\begin{aligned}
x(k) &= x_t(k) + h^{k+1}x_i(k) + h^{N-k+1}x_f(k) \\[2em]
y(k) &= y_t(k)+h^k y_i(k)+h^{N-k+1}y_f(k) \\[2em]
z(k) &= z_t(k) + h^{k+1} z_i(k) + h^{N-k} z_f(k)
\end{aligned}
\right\} \qquad \ldots \quad (3.8)
$$

where

$x_t(k)$, $y_t(k)$ and $z_t(k)$ are the outer series terms,

$x_i(k)$, $y_i(k)$ and $z_i(k)$ are the initial BLC terms, and

$x_f(k)$, $y_f(k)$ and $z_f(k)$ are the final BLC terms.

Substituting (3.8) in (3.1) and separating the outer series, the initial and final BLC series terms, three equations are obtained. The outer series equation is identical with (3.1).

The initial BLC series equation is obtained as

$$
\begin{bmatrix}
hx_i(k+1) \\[1em]
y_i(k+1) \\[1em]
h^2 z_i(k+1)
\end{bmatrix}
=
\begin{bmatrix}
A_{11} & A_{12} & A_{13} \\[1em]
A_{21} & A_{22} & A_{23} \\[1em]
A_{31} & A_{32} & A_{33}
\end{bmatrix}
\begin{bmatrix}
x_i(k) \\[1em]
y_i(k) \\[1em]
z_i(k)
\end{bmatrix}
\qquad \ldots \quad (3.9)
$$

Note that the order of the degenerate model of (3.9) is n_2.

The final BLC series equation is obtained as

$$
\begin{bmatrix} x_f(k+1) \\ y_f(k+1) \\ z_f(k+1) \end{bmatrix} = \begin{bmatrix} hA_{11} & h^2A_{12} & A_{13} \\ hA_{21} & h^2A_{22} & A_{23} \\ hA_{31} & h^2A_{32} & A_{33} \end{bmatrix} \begin{bmatrix} x_f(k) \\ y_f(k) \\ z_f(k) \end{bmatrix} \qquad \dots \quad (3.10)
$$

Note that the degenerate model of (3.10) is of order n_3.

The various series equations for (3.1), (3.9) and (3.10) and the boundary conditions required to solve them are obtained on similar lines given in Chapter 2.

Three-time-scale property

Consider the second order model (3.3) and its eigenvalues p_1, and p_2. For sufficiently small values of h, $|p_1| >> |p_2|$. In the degeneration process as h tends to zero, p_1 approaches infinity and p_2 approaches zero. Thus for this model, p_2 is associated with a stable fast mode which corresponds to a fast decay of solution near k = 0 and p_1 is associated with an unstable fast mode which corresponds to a fast rise near the terminal point k = N. These ideas are easily extended to the $(n_1+n_2+n_3)$ order model (3.1). The n_2 eigenvalues of the stable fast modes have moduli much less than unity which approach zero in the degeneration process. Thus they lead to the formation of the boundary layer at the initial point (k = 0). The n_3 eigenvalues of the unstable fast modes have moduli much larger than unity which approach infinity in the degeneration process. thus they lead to the formation of the boundary layer at the final point (k = N). The n_1 eigenvalues of the slow modes have moduli in the neighbourhood of unity and approach the eigenvalues of $(A_{11} - A_{13} A_{33}^{-1} A_{31})$ in the degeneration process. In the complex plane, the n_1 eigenvalues of the slow modes lie in the neighbourhood of the circumference of the unit circle. The n_2 eigenvalues of the stable fast modes and n_3 eigenvalues of the unstable fast nodes are located near the origin and far outside the circumference of the unit circle respectively. Thus the model (3.1) with three distinct groups of eigenvalues widely separated from each other and contributing to slow, fast decay and fast rise modes is said to possess the three-time-scale property.

Example 3.1 [30]

Consider the fourth order two-point boundary value problem

$$
\begin{bmatrix} x_1(k+1) \\ x_2(k+1) \\ y(k+1) \\ hz(k+1) \end{bmatrix} = \begin{bmatrix} 0 & 1 & 0 & 0 \\ 0 & 0 & 0 & 1 \\ 1 & 0 & 0 & 0 \\ -1.15 & 0.5 & 1.5h & 0.7 \end{bmatrix} \begin{bmatrix} x_1(k) \\ x_2(k) \\ y(k) \\ z(k) \end{bmatrix}
$$

with the boundary conditions

$$
\begin{bmatrix} x_1(0) \\ y(0) \end{bmatrix} = \begin{bmatrix} 15.0 \\ 20.0 \end{bmatrix} \quad ; \quad \begin{bmatrix} x_2(8) \\ z(8) \end{bmatrix} = \begin{bmatrix} 2.0 \\ 3.0 \end{bmatrix}
$$

and the small parameter h = 0.1
The eigenvalues of the above model are
p_1 = - 1.5367 ; p_2 = 0.9291
p_3 = 7.4670 ; p_4 = 0.1407

The eigenvalues p_1 and p_2 correspond to slow modes and p_3 and p_4 correspond to fast rise and fast decay modes respectively.
The eigenvalues of the degenerate model are
p_1 = -1.6877 ; p_2 = 0.9734
which are nearly equal to the eigenvalues of the slow modes.

Using the method developed, the series solutions upto second order approximation are obtained as shown in Table 3.1, and Fig. 3.1. It is noted that the series solutions approach the exact solution with the increased order of approximation.

3.2 Open-Loop Optimal Control Problem [33]

The open-loop optimal control problem of a singularly perturbed, linear, shift invariant discrete-time system in C-model is considered. The resulting two-point boundary value problem (TPBVP) is cast in the singularly perturbed form which exhibits the three-time-

scale property. A singular perturbation method is developed to obtain
the approximate solution composed of an outer series, initial boundary
layer correction series and final boundary layer correction series.

Problem formulation

Consider the linear, shift invariant completely controlable
discrete system

$$
\begin{bmatrix} y(k+1) \\ z(k+1) \end{bmatrix} = \begin{bmatrix} A_{11} & hA_{12} \\ A_{21} & hA_{22} \end{bmatrix} \begin{bmatrix} y(k) \\ z(k) \end{bmatrix} + \begin{bmatrix} B_1 \\ B_2 \end{bmatrix} u(k) \quad \ldots \quad (3.11)
$$

with y(0) and z(0) where
y(k) is (nx1) state vector;
z(k) is (mx1) "   " ;
y(k) is (rx1) control vector

TABLE - 3.1

Comparison of approximate and exact solutions of Example 3.1:

(a) for $x_1(k)$

k	Degenerate solution	Zeroth order solution	Ist order solution	2nd order solution	Exact solution
0	15.0000	15.0000	15.0000	15.0000	15.0000
1	14.3689	14.3689	13.2136	13.1785	13.2061
2	14.3793	14.3793	12.8239	12.5381	12.5138
3	13.3352	13.3352	11.3478	11.1958	11.1269
4	14.0980	14.0980	11.5743	11.1119	11.0872
5	11.8378	11.8378	9.3949	9.2401	9.1458
6	14.7055	14.7055	10.7245	10.2653	10.2680
7	8.9438	8.9438	7.2918	6.8189	6.7927
8	17.7707	17.7707	10.1594	10.2662	10.3428

(b) for $x_2(k)$

k	Degenerate solution	Zeroth order solution	Ist order solution	2nd order solution	Exact solution
0	14.3689	14.3689	13.2136	13.1785	13.2061
1	14.3793	14.3793	12.8239	12.5381	12.5138
2	13.3352	13.3352	11.3478	11.1958	11.1269
3	14.0980	14.0980	11.5743	11.1119	11.0872
4	11.8378	11.8378	9.3949	9.2401	9.1458
5	14.7055	14.7055	10.7245	10.2653	10.2680
6	8.9438	8.9438	7.2918	6.8189	6.7927
7	17.7707	17.7707	10.1594	10.2662	10.3428
8	2.0000	2.0000	2.0000	2.0000	2.0000

TABLE 3.1 (Contd.)
(c) for $y(k)$

k	Degenerate solution	Zeroth order solution	Ist order solution	2nd order solution	Exact solution
0	15.2680	20.0000	20.0000	20.0000	20.0000
1	15.0000	15.0000	15.0000	15.0000	15.0000
2	14.3689	14.3689	13.2136	13.1785	13.2061
3	14.3793	14.3793	12.8239	12.5381	12.5138
4	13.3352	13.3352	11.3478	11.1958	11.1269
5	14.0980	14.0980	11.5743	11.1119	11.0872
6	11.8378	11.8378	9.3949	9.2401	9.1458
7	14.7055	14.7055	10.7245	10.2653	10.2680
8	8.9438	8.9438	7.2918	6.8189	6.7927

(d) for $z(k)$

k	Degenerate solution	Zeroth order solution	Ist order solution	2nd order solution	Exact solution
0	14.3793	14.3793	12.8239	12.5381	12.5138
1	13.3352	13.3352	11.3478	11.1958	11.1268
2	14.0980	14.0980	11.5743	11.1119	11.0872
3	11.8378	11.8378	9.3949	9.2401	9.1458
4	14.7055	14.7055	10.7245	10.2652	10.2680
5	8.9438	8.9438	7.2918	6.8189	6.7927
6	17.7707	17.7707	10.1594	10.2662	10.3428
7	2.0000	2.0000	2.0000	2.0000	2.0000
8	27.7661	3.0000	3.0000	3.0000	3.0000

and h is a small positive scalar aparameter.

A_{ij}, B_i, i, j = 1, 2 are real constant matrices of appropriate dimensionality.

The performance index to be minimized is

$$J = \frac{1}{2} x'(N) S x(N) + \frac{1}{2} \sum_{k=0}^{N-1} [x'(k) D x(k) + u'(k) R u(k)] \qquad \ldots \quad (3.12)$$

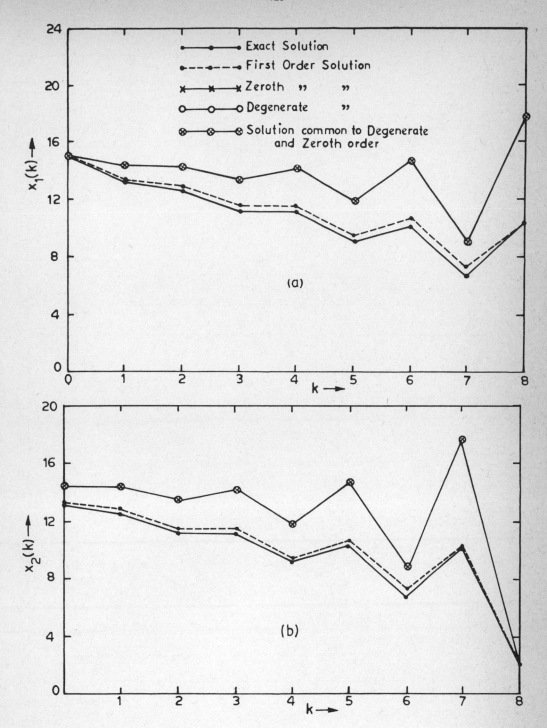

Fig. 3.1(a) & (b) Exact and approximate solutions of $x_1(k)$ & $x_2(k)$ of
Example 3.1

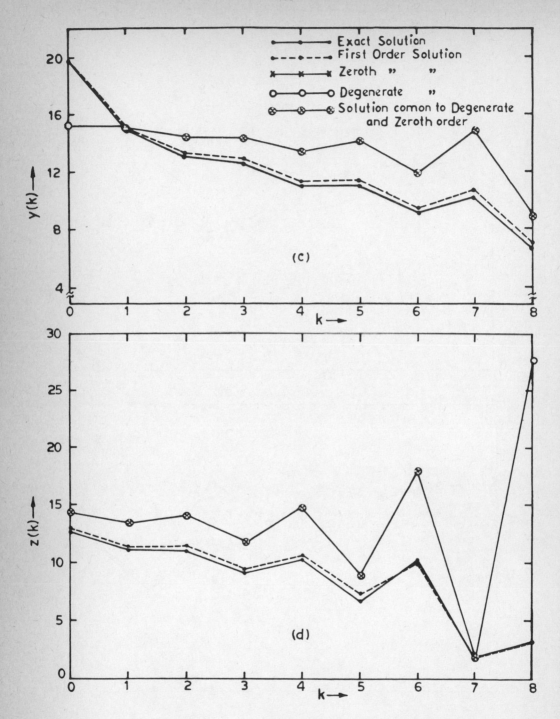

Fig. 3.1(c) & (d)   Exact and approximate solutions of y(k) & z(k) of
Example 3.1

where $x'(k) = [y'(k) \quad hz'(k)]$

D and S are real, positive semidefinite, symmetric matrices of order
$(m+n) \times (m+n)$ given by

$$D = \begin{bmatrix} D_{11} & D_{12} \\ D'_{12} & D_{22} \end{bmatrix} \quad \text{and} \quad S = \begin{bmatrix} S_{11} & S_{12} \\ S'_{12} & S_{22} \end{bmatrix}$$

R is a real, positive definite, symmetric matrix of order $(r \times r)$ N is
a fixed integer indicating the terminal or final value of time.
Note that the states are incorporated in (3.12) in an appropriate
manner as $y(k)$ and $hz(k)$ in order to bring the resulting TPBVP into
singularly perturbed form.

The Hamiltonian of the problem is

$$H(k) = \frac{1}{2} y'(k)D_{11}y(k) + \frac{1}{2} hz'(k)D'_{12} y(k) +$$

$$\frac{1}{2} hy'(k)D_{12}z(k) + \frac{1}{2} h^2 z'(k)D_{22}z(k) + \frac{1}{2} u'(k)Ru(k)$$

$$+p'(k+1) [A_{11}y(k)+hA_{12}z(k)+B_1u(k)]$$

$$+\bar{q}'(k+1) [A_{21}y(k)+hA_{22}z(k)+B_2u(k)] \qquad \cdots \quad (3.13)$$

where $p(k)$ and $\bar{q}(k)$ are the co-states.

Using the results of discrete optimal control theory [41]

$$\frac{\partial H(k)}{\partial p(k+1)} = y(k+1)$$

$$\frac{\partial H(k)}{\partial \bar{q}(k+1)} = z(k+1)$$

$$\frac{\partial H(k)}{\partial y(k)} = p(k) \qquad \cdots \quad (3.14)$$

$$\frac{\partial H(k)}{\partial z(k)} = \bar{q}(k)$$

$$\frac{\partial H(k)}{\partial u(k)} = 0 \quad \text{From (3.13) and (3.14), the state and costate}$$

equations are obtained as

$$
\begin{bmatrix} y(k+1) \\ z(k+1) \\ p(k) \\ q(k) \end{bmatrix} = \begin{bmatrix} A_{11} & hA_{12} & -E_{11} & -hE_{12} \\ A_{21} & hA_{22} & -E'_{12} & -hE_{22} \\ D_{11} & hD_{12} & A'_{11} & hA'_{21} \\ D'_{12} & hD_{22} & A'_{12} & hA'_{22} \end{bmatrix} \begin{bmatrix} y(k) \\ z(k) \\ p(k+1) \\ q(k+1) \end{bmatrix} \qquad \ldots (3.15)
$$

and the optimal control is obtained as

$$
u(k) = -R^{-1} [B'_1 \, p(k+1) + hB'_2 \, q(k+1)] \qquad \ldots (3.16)
$$

where $\bar{q}(k) = hq(k)$; $E_{11} = B_1 R^{-1} B'_1$; $E_{12} = B_1 R^{-1} B'_2$;

$$
E_{22} = B_2 R^{-1} B'_2
$$

The final conditions are

$$
\begin{bmatrix} p(N) \\ q(N) \end{bmatrix} = S \begin{bmatrix} y(N) \\ z(N) \end{bmatrix} \qquad \ldots (3.17)
$$

The equations (3.15) and (3.16) constitute the open-loop optimal control problem. The equation (3.15) is restructured as

$$
\begin{bmatrix} y(k+1) \\ p(k) \\ z(k+1) \\ q(k) \end{bmatrix} = \begin{bmatrix} A_{11} & -E_{11} & hA_{12} & -hE_{12} \\ D_{11} & A'_{11} & hD_{12} & hA'_{21} \\ A_{21} & -E'_{12} & hA_{22} & -hE_{22} \\ D'_{12} & A'_{12} & hD_{22} & hA'_{22} \end{bmatrix} \begin{bmatrix} y(k) \\ p(k+1) \\ z(k) \\ q(k+1) \end{bmatrix} \qquad \ldots (3.18)
$$

The $2(m+n)$th order TPBVP represented by (3.15) or (3.18) is to be solved with the boundary conditions

$$
y(0), \ z(0), \ p(N) \text{ and } q(N).
$$

3.3 Development of Singular Perturbation Method

The 2(m+n)th order TPBVP represented by (3.18) is in the singularly perturbed form in the sense that its degenerate TPBVP

$$
\begin{bmatrix} y^{(0)}(k+1) \\ p^{(0)}(k) \end{bmatrix} = \begin{bmatrix} A_{11} & -E_{11} \\ D_{11} & A'_{11} \end{bmatrix} \begin{bmatrix} y^{(0)}(k) \\ p^{(0)}(k+1) \end{bmatrix} \qquad \ldots \quad (3.19a)
$$

$$
\begin{bmatrix} z^{(0)}(k+1) \\ q^{(0)}(k) \end{bmatrix} = \begin{bmatrix} A'_{21} & -E'_{12} \\ D'_{12} & A'_{12} \end{bmatrix} \begin{bmatrix} y^{(0)}(k) \\ p^{(0)}(k+1) \end{bmatrix} \qquad \ldots \quad (3.19b)
$$

is of reduced order 2m and satisfies the boundary conditions

$$ y^{(0)}(0) = y(0) \quad ; \quad p^{(0)}(N) = p(N) $$

Once (3.19a) is solved with the above boundary conditions, the solutions $z^{(0)}(k)$ and $q^{(0)}(k)$ are fixed from (3.19b) and in general

$$ z^{(0)}(0) \neq z(0) \text{ and } q^{(0)}(N) \neq q(N) $$

The degenerate optimal control is obtained from (3.16) as

$$ u^{(0)}(k) = -R^{-1}B'_1 p^{(0)}(k+1). \qquad \ldots \quad (3.20) $$

Note that (3.18) is in the same form as the model (3.1). It is thus seen that the model (3.18) has the three-time-scale property and two boundary layers are therefore formed at the initial point (k = 0) and the final point (k = N). This situation is similar to that occurring in TPBVP arising in optimal control of continuous systems [43]. Utilizing the results of the analysis in Sec. 3.1, the solution of (3.18) is taken as

$$y(k) = y_t(k) + h^{k+1} y_i(k) + h^{N-k+1} y_f(k)$$

$$p(k) = p_t(k) + h^{k+1} p_i(k) + h^{N-k+1} p_f(k) \qquad \qquad \dots \ (3.21)$$

$$z(k) = z_t(k) + h^k z_i(k) + h^{N-k+1} z_f(k)$$

$$q(k) = q_t(k) + h^{k+1} q_i(k) + h^{N-k} q_f(k)$$

where

$y_t(k)$, $p_t(k)$, $z_t(k)$, $q_t(k)$ correspond to the outer series, $y_i(k)$, $p_i(k)$, $z_i(k)$, $q_i(k)$ correspond to the initial boundary layer correction (BLC) series, and $y_f(k)$, $p_f(k)$, $z_f(k)$, $q_f(k)$ correspond to the final boundary layer correction (BLC) series.

Substituting (3.21) in (3.18) and separating the outer series and the two correction series terms, we get three equations. The outer series equation is found to be the same as (3.18).

The equation for the initial BLC is

$$\begin{bmatrix} hy_i(k+1) \\ z_i(k+1) \\ p_i(k) \\ q_i(k) \end{bmatrix} = \begin{bmatrix} A_{11} & A_{12} & -hE_{11} & -hE_{12} \\ A_{21} & A_{22} & -hE'_{12} & -hE_{22} \\ D_{11} & D_{12} & hA'_{11} & hA'_{21} \\ D'_{12} & D_{22} & hA'_{12} & hA'_{22} \end{bmatrix} \begin{bmatrix} y_i(k) \\ z_i(k) \\ p_i(k+1) \\ hq_i(k+1) \end{bmatrix} \qquad \dots \ (3.22)$$

The initial correction to the optimal control is

$$u_i(k) = -R^{-1}[B'_1 p_i(k+1) + hB'_2 q_i(k+1)]$$

The equation for the <u>final BLC</u> is

$$
\begin{bmatrix}
y_f(k+1) \\
z_f(k+1) \\
hp_f(k) \\
q_f(k)
\end{bmatrix}
=
\begin{bmatrix}
hA_{11} & hA_{12} & -E_{11} & -E_{12} \\
hA_{21} & hA_{22} & -E'_{12} & -E_{22} \\
hD_{11} & hD_{12} & A'_{11} & A'_{21} \\
hD'_{12} & hD_{22} & A'_{12} & A'_{22}
\end{bmatrix}
\begin{bmatrix}
y_f(k) \\
hz_f(k) \\
p_f(k+1) \\
q_f(k+1)
\end{bmatrix}
\qquad \dots \quad (3.23)
$$

The final correction to the optimal control is given by

$$
u_f(k) \;=\; - R^{-1} \,[\, B'_1 p_f(k+1) + hB'_2 \, q_f(k+1) \,]
$$

Series expansions

Assuming series solutions for (3.18), (3.22) and (3.23) and proceeding in the usual way, the series equations for various orders of approximations are obtained as shown

Outer Series:

For zeroth order

$$
\begin{bmatrix}
y^{(0)}(k+1) \\
p^{(0)}(k)
\end{bmatrix}
=
\begin{bmatrix}
A_{11} & -E_{11} \\
D_{11} & A'_{11}
\end{bmatrix}
\begin{bmatrix}
y^{(0)}(k) \\
p^{(0)}(k+1)
\end{bmatrix}
$$

$$
\begin{bmatrix}
z^{(0)}(k+1) \\
q^{(0)}(k)
\end{bmatrix}
=
\begin{bmatrix}
A'_{21} & -E'_{12} \\
D'_{12} & A'_{12}
\end{bmatrix}
\begin{bmatrix}
y^{(0)}(k) \\
p^{(0)}(k+1)
\end{bmatrix}
$$

$$
u^{(0)}(k) \;=\; - R^{-1} B'_1 p^{(0)}(k+1)
$$

For first order

$$
\begin{bmatrix} y^{(1)}(k+1) \\ \\ p^{(1)}(k) \end{bmatrix} = \begin{bmatrix} A_{11} & -E_{11} \\ \\ D_{11} & A'_{11} \end{bmatrix} \begin{bmatrix} y^{(1)}(k) \\ \\ p^{(1)}(k+1) \end{bmatrix} + \begin{bmatrix} A_{12} & -E_{12} \\ \\ D_{12} & A'_{21} \end{bmatrix} \begin{bmatrix} z^{(0)}(k) \\ \\ q^{(0)}(k+1) \end{bmatrix}
$$

$$
\begin{bmatrix} z^{(1)}(k+1) \\ \\ q^{(1)}(k) \end{bmatrix} = \begin{bmatrix} A'_{21} & -E'_{12} \\ \\ D'_{12} & A'_{12} \end{bmatrix} \begin{bmatrix} y^{(1)}(k) \\ \\ p^{(1)}(k+1) \end{bmatrix} + \begin{bmatrix} A_{22} & -E_{22} \\ \\ D_{22} & A'_{22} \end{bmatrix} \begin{bmatrix} z^{(0)}(k) \\ \\ q^{(0)}(k+1) \end{bmatrix}
$$

$$
u^{(1)}(k) = -R^{-1} [B'_1 p^{(1)}(k+1) + B'_2 q^{(0)}(k+1)]
$$

For jth order

$$
\begin{bmatrix} y^{(j)}(k+1) \\ \\ q^{(j)}(k) \end{bmatrix} = \begin{bmatrix} A_{11} & -E_{11} \\ \\ D_{11} & A'_{11} \end{bmatrix} \begin{bmatrix} y^{(j)}(k) \\ \\ p^{(j)}(k+1) \end{bmatrix} + \begin{bmatrix} A_{12} & -E_{12} \\ \\ D_{12} & A'_{21} \end{bmatrix} \begin{bmatrix} z^{(j-1)}(k) \\ \\ q^{(j-1)}(k+1) \end{bmatrix}
$$

$$
\begin{bmatrix} z^{(j)}(k+1) \\ \\ q^{(j)}(k) \end{bmatrix} = \begin{bmatrix} A'_{21} & -E'_{12} \\ \\ D'_{12} & A'_{12} \end{bmatrix} \begin{bmatrix} y^{(j)}(k) \\ \\ p^{(j)}(k+1) \end{bmatrix} + \begin{bmatrix} A_{22} & -E_{22} \\ \\ D_{22} & A'_{22} \end{bmatrix} \begin{bmatrix} z^{(j-1)}(k) \\ \\ q^{(j-1)}(k+1) \end{bmatrix}
$$

$$
\ldots \quad (3.24)
$$

$$
u^{(j)}(k) = -R^{-1} [B'_1 p^{(j)}(k+1) + B'_2 q^{(j-1)}(k-1)]
$$

Initial BLC Series:

For zeroth order

$$
\begin{bmatrix} 0 \\ \\ z_i^{(0)}(k+1) \end{bmatrix}
=
\begin{bmatrix} A_{11} & A_{12} \\ \\ A_{21} & A_{22} \end{bmatrix}
\begin{bmatrix} y_i^{(0)}(k) \\ \\ z_i^{(0)}(k) \end{bmatrix}
$$

$$
\begin{bmatrix} p_i^{(0)}(k) \\ \\ q_i^{(0)}(k) \end{bmatrix}
=
\begin{bmatrix} D_{11} & D_{12} \\ \\ D_{12}' & D_{22} \end{bmatrix}
\begin{bmatrix} y_i^{(0)}(k) \\ \\ z_i^{(0)}(k) \end{bmatrix}
$$

$$
u_i^{(0)}(k) = -R^{-1}B_1' p_i^{(0)}(k+1)
$$

For first order

$$
\begin{bmatrix} y_i^{(0)}(k+1) \\ \\ z_i^{(1)}(k+1) \end{bmatrix}
=
\begin{bmatrix} A_{11} & A_{12} \\ \\ A_{21} & A_{22} \end{bmatrix}
\begin{bmatrix} y_i^{(1)}(k) \\ \\ z_i^{(1)}(k) \end{bmatrix}
+
\begin{bmatrix} -E_{11} & -E_{12} \\ \\ -E_{12}' & -E_{22} \end{bmatrix}
\begin{bmatrix} p_i^{(0)}(k+1) \\ \\ 0 \end{bmatrix}
$$

$$
\begin{bmatrix} p_i^{(1)}(k) \\ \\ q_i^{(1)}(k) \end{bmatrix}
=
\begin{bmatrix} D_{11} & D_{12} \\ \\ D_{12}' & D_{22} \end{bmatrix}
\begin{bmatrix} y_i^{(1)}(k) \\ \\ z_i^{(1)}(k) \end{bmatrix}
+
\begin{bmatrix} A_{11}' & A_{21}' \\ \\ A_{12}' & A_{22}' \end{bmatrix}
\begin{bmatrix} p_i^{(0)}(k+1) \\ \\ 0 \end{bmatrix}
$$

$$
u_i^{(1)}(k) = -R^{-1} [B_1' p_i^{(1)}(k+1) + B_2' q_i^{(0)}(k+1)]
$$

For jth order

$$
\begin{bmatrix} y_i^{(j-1)}(k+1) \\ \\ z_i^{(j)}(k+1) \end{bmatrix} = \begin{bmatrix} A_{11} & A_{12} \\ \\ A_{21} & A_{22} \end{bmatrix} \begin{bmatrix} y_i^{(j)}(k) \\ \\ z_i^{(j)}(k) \end{bmatrix} + \begin{bmatrix} -E_{11} & -E_{12} \\ \\ -E'_{12} & -E_{22} \end{bmatrix} \begin{bmatrix} p_i^{(j-1)}(k+1) \\ \\ q_i^{(j-2)}(k+1) \end{bmatrix}
$$

$$
\begin{bmatrix} p_i^{(j)}(k) \\ \\ q_i^{(j)}(k) \end{bmatrix} = \begin{bmatrix} D_{11} & D_{12} \\ \\ D'_{12} & D_{22} \end{bmatrix} \begin{bmatrix} y_i^{(j)}(k) \\ \\ z_i^{(j)}(k) \end{bmatrix} + \begin{bmatrix} A'_{11} & A'_{21} \\ \\ A'_{12} & A'_{22} \end{bmatrix} \begin{bmatrix} p_i^{(j-1)}(k+1) \\ \\ q_i^{(j-2)}(k+1) \end{bmatrix}
$$

$$\dots \quad (3.25)$$

$$u_i^{(j)}(k) = - R^{-1}[B'_1 p_i^{(j)}(k+1) + B'_2 q_i^{(j-1)}(k+1)]$$

Final BLC Series

For zeroth order

$$
\begin{bmatrix} y_f^{(0)}(k+1) \\ \\ z_f^{(0)}(k+1) \end{bmatrix} = \begin{bmatrix} -E_{11} & -E_{12} \\ \\ -E'_{12} & -E_{22} \end{bmatrix} \begin{bmatrix} p_f^{(0)}(k+1) \\ \\ q_f^{(0)}(k+1) \end{bmatrix}
$$

$$
\begin{bmatrix} 0 \\ \\ q_f^{(0)}(k) \end{bmatrix} = \begin{bmatrix} A'_{11} & A'_{21} \\ \\ A'_{12} & A'_{22} \end{bmatrix} \begin{bmatrix} p_f^{(0)}(k+1) \\ \\ q_f^{(0)}(k+1) \end{bmatrix}
$$

$$u_f^{(0)}(k) = - R^{-1} B'_1 p_f^{(0)}(k+1)$$

For first order

$$
\begin{bmatrix} y_f^{(1)}(k+1) \\ z_f^{(1)}(k+1) \end{bmatrix} = \begin{bmatrix} -E_{11} & -E_{12} \\ -E_{12} & -E_{22} \end{bmatrix} \begin{bmatrix} p_f^{(1)}(k+1) \\ q_f^{(1)}(k+1) \end{bmatrix} + \begin{bmatrix} A_{11} & A_{12} \\ A_{21} & A_{22} \end{bmatrix} \begin{bmatrix} y_f^{(0)}(k) \\ 0 \end{bmatrix}
$$

$$
\begin{bmatrix} p_f^{(0)}(k) \\ q_f^{(1)}(k) \end{bmatrix} = \begin{bmatrix} A'_{11} & A'_{21} \\ A'_{12} & A'_{22} \end{bmatrix} \begin{bmatrix} p_f^{(1)}(k+1) \\ q_f^{(1)}(k+1) \end{bmatrix} + \begin{bmatrix} D_{11} & D_{12} \\ D'_{12} & D_{22} \end{bmatrix} \begin{bmatrix} y_f^{(0)}(k) \\ 0 \end{bmatrix}
$$

$$
u_f^{(1)}(k) = - R^{-1} [B'_1 p_f^{(1)}(k+1) + B'_2 q_f^{(0)}(k+1)]
$$

For the jth order

$$
\begin{bmatrix} y_f^{(j)}(k+1) \\ z_f^{(j)}(k+1) \end{bmatrix} = \begin{bmatrix} -E_{11} & -E_{12} \\ -E'_{12} & -E'_{22} \end{bmatrix} \begin{bmatrix} p_f^{(j)}(k+1) \\ q_f^{(j)}(k+1) \end{bmatrix} + \begin{bmatrix} A_{11} & A_{12} \\ A_{21} & A_{22} \end{bmatrix} \begin{bmatrix} y_f^{(j-1)}(k) \\ z_f^{(j-2)}(k) \end{bmatrix}
$$

$$
\begin{bmatrix} p_f^{(j-1)}(k) \\ q_f^{(j)}(k) \end{bmatrix} = \begin{bmatrix} A'_{11} & A'_{21} \\ A'_{12} & A'_{22} \end{bmatrix} \begin{bmatrix} p_f^{(j)}(k+1) \\ q_f^{(j)}(k+1) \end{bmatrix} + \begin{bmatrix} D_{11} & D_{12} \\ D'_{12} & D_{22} \end{bmatrix} \begin{bmatrix} y_f^{(j-1)}(k) \\ z_f^{(j-2)}(k) \end{bmatrix}
$$

$$
\ldots \quad (3.26)
$$

$$
u_f^{(j)}(k) = - R^{-1} [B'_1 p_f^{(j)}(k+1) + B'_2 q_f^{(j-1)}(k+1)]
$$

Total series solution and evaluation of boundary conditions of boundary conditions series equations

The total series solution is composed of the outer series, initial BLC series and the final BLC series. That is,

$$
y(k,h) = \sum_{r=0}^{j} [y^{(r)}(k) + h^{k+1} y_i^{(r)}(k) + h^{N-k+1} y_f^{(r)}(k)] h^r
$$

$$
z(k,h) = \sum_{r=0}^{j} [z^{(r)}(k) + h^k z_i^{(r)}(k) + h^{N-k+1} z_f^{(r)}(k)] h^r
$$

$$
\left. \qquad \qquad \qquad \qquad \qquad \qquad \qquad \qquad \qquad \right\} \quad \dots \quad (3.27)
$$

$$
p(k,h) = \sum_{r=0}^{j} [p^{(r)}(k) + h^{k+1} p_i^{(r)}(k) + h^{N-k+1} p_f^{(r)}(k)] h^r
$$

$$
q(k,h) = \sum_{r=0}^{j} [q^{(r)}(k) + h^{k+1} q_i^{(r)}(k) + h^{N-k} q_f^{(r)}(k)] h^r
$$

The boundary conditions required to solve the series equations (3.24) to (3.26) are easily obtained from (3.27). They are

$$
h^0 : \quad y^{(0)}(0) = y(0) \ ; \ p^{(0)}(N) = p(N)
$$

$$
z_i^{(0)}(0) = z(0) - z^{(0)}(0) \ ; \ q_f^{(0)}(N) = q(N) - q^{(0)}(N)
$$

$$
h^1 : \quad y^{(1)}(0) = -y_i^{(0)}(0) \ ; \ p^{(1)}(N) = -p_f^{(0)}(N)
$$

$$
z_i^{(1)}(0) = -z^{(0)}(0) \ ; \ q_f^{(1)}(N) = -q^{(1)}(N)
$$

$$
h^j : \quad y^{(j)}(0) = -y_i^{(j-1)}(0) \ ; \ p^{(j)}(N) = -p_f^{(j-1)}(N)
$$

$$
z_i^{(j)}(0) = -z^{(j)}(0) \ ; \ q_f^{(j)}(N) = -q^{(j)}(N)
$$

Example 3.2 [33]

Consider the second order control system

$$
\begin{bmatrix} y(k+1) \\ \\ z(k+1) \end{bmatrix} = \begin{bmatrix} 0.95 & h \\ \\ 1 & 0 \end{bmatrix} \begin{bmatrix} y(k) \\ \\ z(k) \end{bmatrix} + \begin{bmatrix} 1 \\ \\ 0 \end{bmatrix} u(k) \quad \dots \quad (3.28)
$$

where the small parameter h = 0.12; the initial conditions

$$y(0) = 10.0 \; ; \; z(0) = 15.0$$

and the performance index

$$J = \frac{1}{2} \sum_{k=0}^{N-1} [x'(k) \; Dx(k) + R \; u(k)^2] \qquad \cdots \quad (3.29)$$

with

$$D = \begin{bmatrix} 1 & 0 \\ 0 & 2 \end{bmatrix} ; \quad x(k) = \begin{bmatrix} y(k) \\ hz(k) \end{bmatrix} ; \quad R = 2 \text{ and } N = 8$$

The eigenvalues of (3.27) are

$$p_1 = 0.8000 \; ; \; p_2 = 0.1500$$

The singularly perturbed TPBVP of fourth order corresponding to (3.15) is

$$\begin{bmatrix} y(k+1) \\ z(k+1) \\ p(k) \\ q(k) \end{bmatrix} = \begin{bmatrix} 0.95 & h & -0.5 & 0 \\ 1.0 & 0 & 0 & 0 \\ 1.0 & 0 & 0.95 & h \\ 0 & 2.0h & 1.0 & 0 \end{bmatrix} \begin{bmatrix} y(k) \\ z(k) \\ p(k+1) \\ q(k+1) \end{bmatrix} \qquad \cdots \quad (3.30)$$

and the optimal control is given by

$$u(k) = - 0.5 \; p(k+1)$$

The boundary conditions for solving (3.30) are

$$\begin{bmatrix} y(0) \\ z(0) \end{bmatrix} = \begin{bmatrix} 10.0 \\ 15.0 \end{bmatrix} ; \quad \begin{bmatrix} p(N) \\ q(N) \end{bmatrix} = \begin{bmatrix} 0 \\ 0 \end{bmatrix}$$

The eigenvalues of (3.30) are

$$p_1 = 1.8370 \quad ; \quad p_2 = 0.5444$$
$$p_3 = -9.2398 \quad ; \quad p_4 = -0.1082$$

The eigenvalues of the degenerate model of (3.30) are

$$p_1 = 2.0383 \quad ; \quad p_2 = 0.4906$$

which are nearly equal to the eigenvalues of the slow modes given above.

Using the singular perturbation method developed, the degenerate, zeroth, first, and second order solutions are evaluated and compared with the exact solution as shown in Table 3.2 and Fig. 3.2.

Using the control u(k), corresponding to exact, degenerate, zeroth, first and second order solutions, the resulting states y(k) and z(k) are calculated from the original system of equations (3.28) and the corresponding performance index is evaluated from (3.29). The results are shown in Table 3.3.

Note: The exact solution of the fourth order singularly perturbed discrete TPBVP given by (3.30) is obtained using the method of complementary functuions suggested for continuous "stiff" problems [44].

3.4 Conclusions [30,33]

A state space model having the three-time-scale property and characterised by three distinct groups of eigenvalues has been formulated. In the process of degeneration, it is found that this model loses some of the initial and final conditions. A method of getting the approximate solution in terms of the outer series, the initial boundary layer correction series and final boundary layer correction series has been developed.

The open-loop optimal control problem with a quadratic performance index has been considered. Using the discrete maximum principle, the state and costate equations are obtained and cast in the singularly perturbed form having the three-time-scale property. A mehtod has been proposed to solve the resulting singularly perturbed TPBVP.

The open-loop discrete optimal control problem exhibiting the three-time-scale property has the same essential features as its counterpart in continuous systems [43].

TABLE - 3.2

Comparison of approximate and exact (optimal) solutions of Example 3.2

k	Degenerate solution	Zeroth order solution	Ist order solution	2nd order solution	Exact solution
y(0)	10.0000	10.0000	10.0000	10.0000	10.0000
z(0)	20.3833	15.0000	15.0000	15.0000	15.0000
p(0)	18.7282	18.7282	21.8053	22.5680	22.5815
q(0)	9.1876	9.1876	15.4574	15.6991	15.7059
u(0)	-4.5938	-4.5938	-5.9287	-6.0496	-6.0529
y(1)	4.9062	4.9062	5.3713	5.2504	5.2471
z(1)	10.0000	10.0000	10.0000	10.0000	10.0000
p(1)	9.1876	9.1876	11.8574	12.0991	12.1059
q(1)	4.5068	4.5067	8.9483	9.0096	9.0068
u(1)	-2.2533	-2.2533	-3.2742	-3.3048	-3.3037
y(2)	2.4159	2.4159	3.0369	2.8914	2.8810
z(2)	4.9062	4.9062	5.3713	5.2504	5.2471
p(2)	4.5067	4.5067	6.5483	6.6096	6.6073
q(2)	2.2096	2.2096	4.7459	4.8960	4.8487
u(2)	-1.1048	-1.1048	-1.7842	-1.8035	-1.7947
y(3)	1.1824	1.1824	1.6817	1.5801	1.5719
z(3)	2.4159	2.4159	3.0369	2.8914	2.8810
p(3)	2.2096	2.2096	3.5685	3.6069	3.5894
q(3)	1.0812	1.0812	2.4996	2.6985	2.6360
u(3)	-0.5406	-0.5406	-0.9599	-0.9848	-0.9723
y(4)	0.5827	0.5827	0.9276	0.8807	0.8668
z(4)	1.1824	1.1824	1.6817	1.5801	1.5720
p(4)	1.0812	1.0812	1.9198	1.9697	1.9445
q(4)	0.5248	0.5248	1.2972	1.4678	1.4180
u(4)	-0.2624	-0.2624	-0.5067	-0.5268	-0.5204
y(5)	0.2912	0.2912	0.5165	0.5064	0.4918
z(5)	0.5827	0.5927	0.9276	0.8807	0.8668
p(5)	0.5248	0.5248	1.0134	1.0536	1.0407
q(5)	0.2460	0.2460	0.6506	0.7737	0.7417
u(5)	-0.1230	-0.1230	-0.2554	-0.2755	-0.2668
y(6)	0.1537	0.1537	0.3053	0.3170	0.3044
z(6)	0.2912	0.2912	0.5165	0.5064	0.4918
p(6)	0.2460	0.2460	0.5108	0.5510	0.5336
q(6)	0.0973	0.0973	0.2864	0.3643	0.3501
u)6)	-0.0486	-0.0486	-0.1082	-0.1210	-0.1161
y(7)	0.0976	0.0976	0.2169	0.2423	0.2321
z(7)	0.1537	0.1537	0.3053	0.3170	0.3044
p(7)	0.0973	0.0973	0.2165	0.2419	0.2321
q(7)	0.0000	0.0000	0.0368	0.0732	0.0730
u(7)	0.0000	0.0000	0.0000	0.0000	0.0000

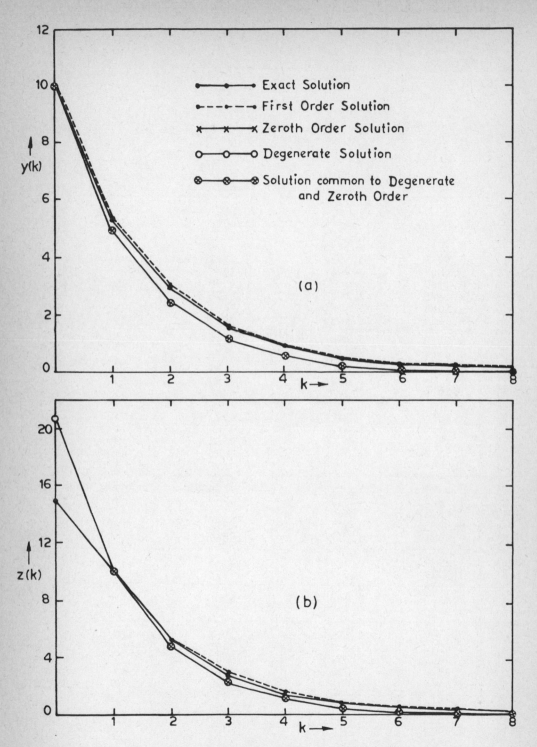

Fig. 3.2(a) & (b)   Exact and approximate solutions of y(k) & z(k) of
Example 3.2

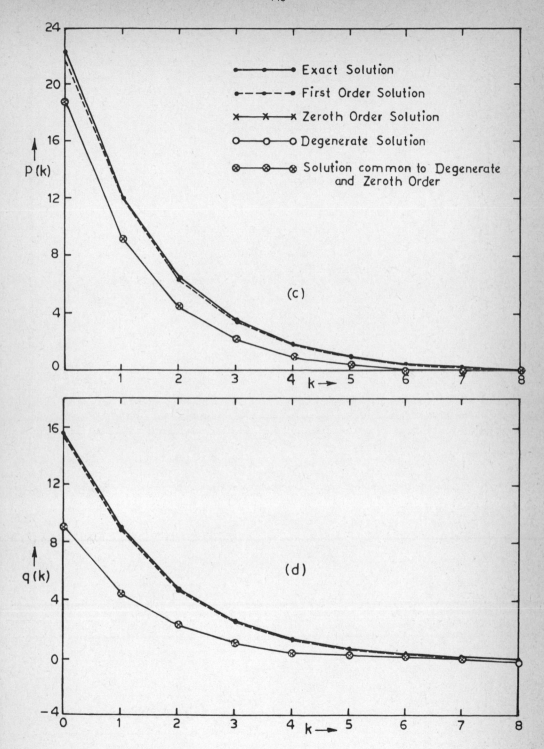

Fig. 3.2(c) & (d) Exact and approximate solutions of p(k) & q(k) of
Example 3.2

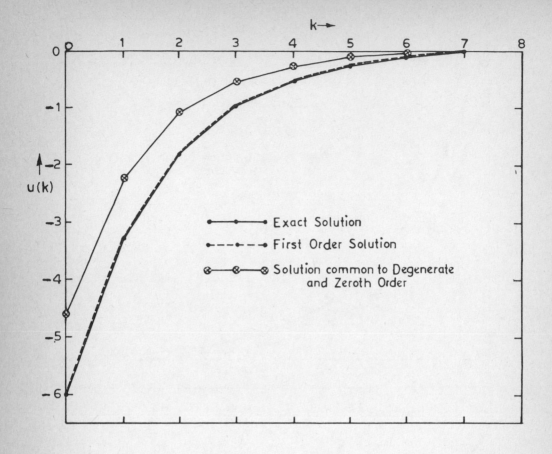

Fig. 3.2(e)   Exact and approximate solutions of u(k) of Example 3.2

TABLE 3.2 (Contd.)

k	Degenerate solution	Zeroth order solution	Ist order solution	2nd order solution	Exact solution
y(8)	0.0931	0.0931	0.2249	0.2672	0.2570
z(8)	0.0976	0.0976	0.2169	0.2423	0.2321
p(8)	0.0000	0.0000	0.0000	0.0000	0.0000
q(8)	-0.0973	0.0000	0.0000	0.0000	0.0000

TABLE - 3.3

Comparison of performance indices

Description of solution	Performance Index
Exact (optimal solution)	127.0424
Degenerate solution	179.5604
Zeroth order solution	179.5604
First order solution	127.1949
Second order solution	127.0448

CHAPTER - 4

SINGULARLY PERTURBED NONLINEAR DIFFERENCE EQUATIONS AND

CLOSED-LOOP DISCRETE OPTIMAL CONTROL PROBLEM

In this Chapter, first a method is proposed to analyze the
singularly perturbed nonlinear difference equations for initial and
boundary value problems. The approximate solution is obtained in the
form of an outer series and a correction series. It is found that
considerable care has to be taken in formulating the equations for the
boundary layer correction series in the case of nonlinear equations.
Then the closed-loop optimal control problem is formulated resulting in
the singularly perturbed nonlinear matrix Riccati difference equation. It is seen that the degeneration (the process of suppressing a
small parameter) affects some of the final conditions of the Riccati
equation. A method is developed to obtain approximate solution in
terms of an outer series and a terminal boundary layer correction
series. A method is also proposed for the important case of the steady
state solution of the matrix Riccati equation. It is found that the
proposed mehtods with the special feature of order reduction, offers
considerable computational simplicity in evaluating the inverse of a
matrix associated with the solution of Riccati equation. Examples are
given to illustrate the proposed methods [34].

4.1 Nonlinear Difference Equations

Consider a second order nonlinear difference equation

$$y(k+2) + ay(k+1) + hy^2(k) = 0 \qquad \qquad \dots \quad (4.1)$$

where h is the small parameter.

Letting h = 0 in (4.1), the degenerate equation

$$y^{(0)}(k+2) + ay^{(0)}(k+1) = 0$$

is of reduced order, and thus (4.1) is said to be in the singularly
perturbed form. There can be many other types of singularly perturbed
nonlinear difference equations.

Case (a):   Initial Value Problem

For (4.1), the given initial conditions are y(0) and y(1). Since
(4.1) belongs to R-type equation dealt in detail in Chapter 1, the
degenerate equation of (4.1) satisfies y(1) only and loses the other
initial condition y(0). The boundary layer occurs at k = 0. A
correction is incorporated by using the transformation

$$w(k) = y(k)/h^k$$

The total solution is given by

$$y(k) = y_t(k) + h^k w(k) \qquad \dots \quad (4.2)$$

where $y_t(k)$ refers to the outer solution and w(k) refers to the
boundary layer correction solution.

Using (4.2) in (4.1)

$$[y_t(k+2)+h^{k+2}w(k+2)] + a\,[y_t(k+1)+h^{K+1}w(k+1)] + h[y_t(k)+h^k w(k)]^2 = 0$$

From the above equation, separating the outer series terms, we get

$$y_t(k+2) + ay_t(k+1) + hy_t^2(k) = 0 \qquad \dots \quad (4.3)$$

The remaining terms for the correction series are obtained as

$$hw(k+2) + aw(k+1) + h^k w^2(k) + 2y_t(k)w(k) = 0 \qquad \dots \quad (4.4)$$

Note that (4.3) is identical to (4.1) and therefore (4.1) is as well
refered to as the equation of the outer series.

Outer Series:

Assume the outer series solution as

$$y(k) = y^{(0)}(k) + hy^{(1)}(k) + h^2 y^{(2)}(k) + \ldots \ldots$$

Substituting the above solution in (4.1) and equating terms with like powers of h on either side, we get the following equations

$$h^0 \; : \quad y^{(0)}(k+2) + ay^{(0)}(k+1) = 0$$

$$h^1 \; : \quad y^{(1)}(k+2) + ay^{(1)}(k+1) = - y^{(0)2}(k)$$

$$h^2 \; : \quad y^{(2)}(k+2) + ay^{(2)}(k+1) = -2y^{(0)}(k)y^{(1)}(k)$$

Equations for higher order approximations are obtained similarly.

Correction Series:

Assume the correction series solution as

$$w(k) = w^{(0)}(k) + hw^{(1)}(k) + h^2w^{(2)}(k) + \ldots \ldots$$

By substitution of the above in (4.4) and comparison of coefficients, we get the equations corresponding to zeroth, first and higher order approximations. These equations require careful handling in view of the terms h^k in (4.4). However, since the correction series need to be evaluated for only few values of k, say k = 0 and 1, it does not pose any serious problems as shown below.

Substituting k = 0 for h^k in (4.4),

$$hw(k+2) + aw(k+1) + w^2(k) + 2y(k) \, w(k) = 0 \qquad \ldots \quad (4.5a)$$

For the above equation, the zeroth order approximation is

$$aw^{(0)}(k+1) + w^{(0)2}(k) + 2y^{(0)}(k) \, w^{(0)}(k) = 0 \qquad \ldots \quad (4.5b)$$

Using k = 1 for h^k in (4.4),

$$hw(k+2) + aw(k+1) + hw^2(k) + 2y(k)w(k) = 0 \qquad \ldots \quad (4.5c)$$

For the above equation, the zeroth order approximation is

$$aw^{(0)}(k+1) + 2y^{(0)}(k) \, w^{(0)}(k) = 0 \qquad \ldots \quad (4.5d)$$

Thus (4.5b) and (4.5d) are the two equations for the zeroth order approximation only and should be used only for $k = 0$ and $k = 1$ respectively.

For first order approximation, (4.5a) and (4.5c) become for $k = 0$,

$$aw^{(1)}(k+1) + 2w^{(0)}(k) \, w^{(1)}(k) + 2y^{(0)}(k) \, w^{(1)}(k) =$$

$$- 2y^{(1)}(k)w^{(0)}(k) - w^{(0)}(k+2) \quad \dots \quad (4.5e)$$

for $k = 1$

$$aw^{(1)}(k+1) + 2y^{(0)}(k)w^{(1)}(k) = - w^{(0)2}(k) - 2y^{(1)}(k)w^{(0)}(k)$$

$$- w^{(0)}(k+2) \quad \dots \quad (4.5f)$$

Similar equations are obtained for higher order approximations and for higher values of k.

Total series solution:

The total series solution consisting of the outer series solution and correction series solution is given by

$$y(k) = (y^{(0)}(k) + h \, y^{(1)}(k) + h^2 y^{(2)}(k) + \ldots \ldots)$$

$$+ h^k (w^{(0)}(k) + hw^{(1)}(k) + \ldots \ldots) \quad \dots \quad (4.6)$$

The initial conditions required for solving the series equations (4.5) are obtained from the above equation as

$$h^0 : \quad y^{(0)}(1) = y(1) \; ; \; w^{(0)}(0) = y(0) - y^{(0)}(0)$$

$$h^1 : \quad y^{(1)}(1) = - w^{(0)}(1) \; ; \; w^{(1)}(0) = - y^{(1)}(0)$$

$$\cdot \, \cdot \quad \cdot \cdot \cdot \cdot \cdot \cdot \cdot \cdot \cdot \cdot \cdot \cdot \cdot \cdot \cdot \cdot \cdot \cdot$$

$$h^j : \quad y^{(j)}(1) = - w^{(j-1)}(1) \; ; \; w^{(j)}(0) = - y^{(j)}(0)$$

It is to be noted that for a desired second order approximation, the correction series terms $w^{(0)}(1)$, $w^{(0)}(2)$, and $w^{(1)}(1)$ are obtained by solving the corresponding correction series equations (4.5b),

(4.5d), and (4.5e) respectively. On the otherhand terms $w^{(0)}(0)$ and $w^{(1)}(0)$ are simply evaluated from the outer series coefficients $y(0) - y^{(0)}(0)$ and $y^{(1)}(0)$ respectively.

Example 4.1

Consider the second order nonlinear equation

$$y(k+2) - 0.75y(k+1) + hy(k)^2 = 0$$

with the initial conditions

$$y(0) = 1.0 \; ; \; y(1) = 2.0 \text{ and } h = 0.05$$

Using the method developed above the degenerate, zeroth, first and second order series equations are obtained and solved. The results are shown in Table 4.1 which indicate that the series solutions approach the exact solution with the increased order of approximation.

TABLE - 4.1

Comparison of approximate and exact solutions of
Example 4.1

y(k)	Degenerate solution	Zeroth order solution	First order solution	Second order solution	Exact solution
y(0)	2.6667	1.0000	1.0000	1.0000	1.0000
y(1)	2.0000	2.0000	2.0000	2.0000	2.0000
y(2)	1.5000	1.5000	1.4500	1.4500	1.4500
y(3)	1.1250	1.1250	0.8875	0.8875	0.8875
y(4)	0.8438	0.8438	0.5531	0.5606	0.5605
y(5)	0.6328	0.6328	0.3516	0.3839	0.3810
y(6)	0.4746	0.4746	0.2281	0.2769	0.2700
y(7)	0.3560	0.3560	0.1510	0.2054	0.1953
y(8)	0.2670	0.2670	0.1020	0.1545	0.1429
y(9)	0.2002	0.2002	0.0701	0.1168	0.1052
y(10)	0.1502	0.1502	0.0491	0.0885	0.0779

Note: The degenerate and zeroth order solution are the same except at k = 0

Case (b): Boundary Value Problem

For the BVP, given $y(0)$ and $y(N)$, the procedure is almost similar to the case of IVP. The selection of the boundary condition is however different and is furnished below.

h^0 : $y^{(0)}(N) = y(N)$; $w^{(0)}(0) = y(0) - y^{(0)}(0)$

h^1 : $y^{(1)}(N) = 0$; $w^{(1)}(0) = - y^{(1)}(0)$

. . . .

h^j : $y^{(j)}(N) = 0$; $w^{(j)}(0) = - y^{(j)}(0)$

4.2 Closed-Loop Optimal Control Problem

Consider a linear, shift-invariant, completely controllable singularly perturbed discrete system described by

$$
\begin{bmatrix} y(k+1) \\ z(k+1) \end{bmatrix} = \begin{bmatrix} A_{11} & hA_{12} \\ A_{21} & hA_{22} \end{bmatrix} \begin{bmatrix} y(k) \\ z(k) \end{bmatrix} + \begin{bmatrix} B_1 \\ B_2 \end{bmatrix} u(k) \qquad \ldots \quad (4.7)
$$

with initial conditions

$$
y(k=0) = y(0) ; z(k=0) = z(0)
$$

where

$y(k)$ and $z(k)$ are $(n_1 \times 1)$ and $(n_2 \times 1)$ state vectors respectively; $u(k)$ is an $(r \times 1)$ control vector;

A_{ij}, B_i, i, j = 1, 2 are real constant matrices of compatible dimensionality.

The performance index to be minimized is

$$
J = \frac{1}{2} x'(N)Sx(N) + \frac{1}{2} \sum_{k=0}^{N-1} [x'(k) \, Qx(k) + u'(k) \, Ru(k)] \qquad \ldots \quad (4.8)
$$

where

$x'(k) = (y'(k), z'(k));$

S and Q are real, positive semidefinite $(n_1+n_2) \times (n_1+n_2)$ order symmetric matrices;

R is a real, positive definite $(r \times r)$ order symmetric matrix.

Using the well-known results of optimal control theory [41], the closed-loop optimal control is given by

$$u(k) = - R^{-1}B' [P^{-1}(k+1) + BR^{-1}B']^{-1} A\ x(k)$$

$$= - R^{-1}B'P(k+1) [I + BR^{-1}B'P(k+1)]^{-1} Ax(k) \qquad \ldots \quad (4.9)$$

where the positive definite symmetric matrix P(k+1) of order $(n_1+n_2) \times (n_1+n_2)$ satisfies the matrix Riccati difference equation

$$P(k) = Q + A'P(k+1) [I + BR^{-1}B'P(k+1)]^{-1} A \qquad \ldots \quad (4.10)$$

with the final condition P(N) = S

$$A = \begin{bmatrix} A_{11} & hA_{12} \\ A_{21} & hA_{22} \end{bmatrix} \quad ; \quad B = \begin{bmatrix} B_1 \\ B_2 \end{bmatrix} \quad ; \quad Q = \begin{bmatrix} Q_1 & hQ_2 \\ hQ_2' & h^2Q_3 \end{bmatrix}$$

$$P(k) = \begin{bmatrix} P_1(k) & hP_2(k) \\ hP_2'(k) & h^2P_3(k) \end{bmatrix} \quad ; \quad S = \begin{bmatrix} S_1 & hS_2 \\ hS_2' & h^2S_3 \end{bmatrix} \qquad \ldots \quad (4.11)$$

The Riccati equation (4.10) is solved backward in time with the condition P(N) = S and certain gain functions called "Kalman gains" are obtained. These precomputed gains are stored and applied to the system (4.7) as it runs forward in time. Thus we have a closed-loop optimal control problem.

Substituting (4.11) in (4.10) and rearranging (Appendix 4.1)

$$P_1(k) = Q_1 + A_{11}'(F_1(k+1)A_{11} + F_2(k+1)A_{21}) + A_{21}'(F_2'(k+1)A_{11} + F_3(k+1)A_{21})$$

$$\cdots \quad (4.12a)$$

$$P_2(k) = Q_2 + A_{11}'(F_1(k+1)A_{12} + F_2(k+1)A_{22}) + A_{21}'(F_2'(k+1)A_{12} + F_3(k+1)A_{22})$$

$$\cdots \quad (4.12b)$$

$$P_3(k) = Q_3 + A_{12}'(F_1(k+1)A_{12} + F_2(k+1)A_{22}) + A_{22}'(F_2'(k+1)A_{12} + F_3(k+1)A_{22})$$

$$\cdots \quad (4.12c)$$

where

$$\begin{bmatrix} F_1(k+1) & F_2(k+1) \\ \\ F_2'(k+1) & F_3(k+1) \end{bmatrix} =$$

$$\begin{bmatrix} P_1(k+1) & hP_2(k+1) \\ \\ hP_2'(k+1) & h^2P_3(k+1) \end{bmatrix} \left[\left\{ \begin{array}{cc} I_1 & 0 \\ \\ 0 & I_2 \end{array} \right\} + \left\{ \begin{array}{cc} E_1 & E_2 \\ \\ E_2' & E_3 \end{array} \right\} \left\{ \begin{array}{cc} P_1(k+1) & hP_2(k+1) \\ \\ hP_2'(k+1) & h^2P_3(k+1) \end{array} \right\} \right]^{-1}$$

$$\cdots \quad (4.13)$$

$$\begin{bmatrix} E_1 & E_2 \\ \\ E_2' & E_3 \end{bmatrix} = \begin{bmatrix} B_1 R^{-1} B_1' & B_1 R^{-1} B_2' \\ \\ B_2 R^{-1} B_1' & B_2 R^{-1} B_2' \end{bmatrix}$$

Using h = 0 in (4.13),

$$\begin{bmatrix} F_1^{(0)}(k+1) & F_2^{(0)}(k+1) \\ \\ F_2^{(0)'}(k+1) & F_3^{(0)}(k+1) \end{bmatrix} = \begin{bmatrix} P_1^{(0)}(k+1) \{I_1 + E_1 P_1^{(0)}(k+1)\}^{-1} & 0 \\ \\ 0 & 0 \end{bmatrix}$$

$$\cdots\cdots \quad (4.14)$$

Substituting (4.14) in (4.12), the degenerate matrix Riccati equation becomes

$$P_1^{(0)}(k) = Q_1 + A_{11}' F_1^{(0)}(k+1) A_{12} \qquad \qquad \ldots \ (4.15a)$$

$$P_2^{(0)}(k) = Q_2 + A_{11}' F_1^{(0)}(k+1) A_{12} \qquad \qquad \ldots \ (4.15b)$$

$$P_3^{(0)}(k) = Q_3 + A_{12}' F_1^{(0)}(k+1) A_{12} \qquad \qquad \ldots \ (4.15c)$$

Noting from (4.14) that $F_1^{(0)}(k+1)$ contains $P_1^{(0)}(k+1)$ only, it is seen that (4.15a) is a difference equation of order n_1 whereas (4.15b) and (4.15c) are algebraic equations. That is, if (4.15a) is solved with

$$P_1^{(0)}(N) = P_1(N) = S_1 \qquad \qquad \ldots \ (4.16a)$$

The solutions for $P_2^{(0)}(k)$ and $P_3^{(0)}(k)$ are automatically fixed from (4.15b) and (4.15c) respectively. This results in

$$P_2^{(0)}(N) \neq P_2(N) \ ; \ P_3^{(0)}(N) \neq P_3(N) \qquad \qquad \ldots \ (4.16b)$$

Thus in the degeneration process, the (n_1+n_2) order matrix Riccati equation (4.12) is reduced to n_1 order matrix Riccati equation (4.15) and in the process lost the final conditions (4.16b). Thus (4.12) is said to be in the singularly perturbed form.

Note: It is seen that the solution of the original matrix Riccati equation (4.12) invloves the inverse of the matrix (4.13) of order (n_1+n_2) whereas the degenerate matrix Riccati equation (4.15) requires the inverse of the matrix (4.14) of order n_1.

4.3 Boundary Layer Correction (BLC)

The degenerate or reduced equation (4.15) offers the advantage of order reduction, but suffers from the inability to satisfy all the given final conditions. This loss of some of the final conditions (4.16b) is attributed to the existence of the boundary layer corresponding to the final point $k = N$. In order to recover these final conditions that are lost in the process of degeneration, a boundary layer correction is incorporated by using certain transformations as explained below.

The total solution $P(k)$ is written as

$$P(k) = \bar{P}(k) + \tilde{W}(k) \qquad \qquad \cdots \quad (4.17)$$

where,

$\bar{P}(k)$ refers to outer solution and $\tilde{W}(k)$ refers to boundary layer correction solution.

$$\bar{P}(k) = \begin{bmatrix} \bar{P}(k) & h\bar{P}_2(k) \\ \\ h\bar{P}_2'(k) & h^2\bar{P}_3(k) \end{bmatrix} \qquad \tilde{W}(k) = \begin{bmatrix} \tilde{W}_1(k) & h\tilde{W}_2(k) \\ \\ h\tilde{W}_2'(k) & h^2\tilde{W}_3(k) \end{bmatrix}$$

Based on the results of Chapter 2, where the boundary layer occurs at the final point $k = N$, the transformations are given by

$$W_1(k) = \tilde{W}_1(k)/h^{N-k+1} ; \ W_2(k) = \tilde{W}_2(k)/h^{N-k} ; \ W_3(k) = \tilde{W}_3(k)/h^{N-k}$$

$$\cdots \quad (4.18a)$$

Therefore,

$$\tilde{W}(k) = h^{N-k+1} \begin{bmatrix} W_1(k) & W_2(k) \\ \\ W_2'(k) & hW_3(k) \end{bmatrix} = h^{N-k+1}W(k) \qquad \cdots \quad (4.18b)$$

$$\tilde{W}(k+1) = h^{N-k}W(k+1) \qquad \qquad \cdots \quad (4.18c)$$

Using the above, (4.17) becomes

$$P(k+1) = \bar{P}(k+1) + h^{N-k}W(k+1) \qquad \cdots \quad (4.19)$$

Substituting (4.19) in (4.13) and omitting the argument $(k+1)$ for simplicity, we get

$$F = [\bar{P} + h^{N-k} \ W] \ [I + E \ \bar{P} + h^{N-k} \ EW]^{-1} \qquad \cdots \quad (4.20a)$$

Rearranging the terms (Appendix 4.2), we get

$$F = \bar{F} + h^{N-k}G \qquad \qquad \cdots \quad (4.20b)$$

where,

$$\bar{F} = \bar{P} C \;;\; C = [I + E \bar{P}]^{-1}$$

$$G = WC - \bar{P} C E W C - h^{N-k} WCEWC;$$

Substituting (4.20b) and (4.19) in (4.12), we get

$$
\begin{bmatrix}
\bar{P}_1(k) + h^{N-k+1}W_1(k) & \bar{P}_2(k) + h^{N-k}W_2(k) \\[2ex]
\bar{P}_2'(k) + h^{N-k}W_2'(k) & \bar{P}_3(k) + h^{N-k}W_3(k)
\end{bmatrix}
$$

$$
=
\begin{bmatrix}
Q_1 & Q_2 \\[2ex]
Q_2' & Q_3
\end{bmatrix}
+
\begin{bmatrix}
A_{11}' & A_{21}' \\[2ex]
A_{12}' & A_{22}'
\end{bmatrix}
\left\{
\begin{bmatrix}
\bar{F}_1(k+1) & \bar{F}_2(k+1) \\[2ex]
\bar{F}_2'(k+1) & \bar{F}_3(k+1)
\end{bmatrix}
\right\}
$$

$$
+ h^{N-k}
\left\{
\begin{bmatrix}
G_1(k+1) & G_2(k+1) \\[2ex]
G_2'(k+1) & G_3(k+1)
\end{bmatrix}
\right\}
\begin{bmatrix}
A_{11} & A_{12} \\[2ex]
A_{21} & A_{22}
\end{bmatrix}
\quad \ldots \;(4.21)
$$

Separating the outer series terms from (4.21)

$$
\begin{bmatrix}
\bar{P}_1(k) & \bar{P}_2(k) \\[2ex]
\bar{P}_2'(k) & \bar{P}_3(k)
\end{bmatrix}
=
\begin{bmatrix}
Q_1 & Q_2 \\[2ex]
Q_2' & Q_3
\end{bmatrix}
+
\begin{bmatrix}
A_{11}' & A_{21}' \\[2ex]
A_{12}' & A_{22}'
\end{bmatrix}
\begin{bmatrix}
\bar{F}_1(k+1) & \bar{F}_2(k+1) \\[2ex]
\bar{F}_2'(k+1) & \bar{F}_3(k+1)
\end{bmatrix}
\begin{bmatrix}
A_{11} & A_{12} \\[2ex]
A_{21} & A_{22}
\end{bmatrix}
$$

$$\ldots \;(4.22)$$

Separating the correction series terms from (4.21) and dividing all the terms by h^{N-k}

$$
\begin{bmatrix}
hW_1(k) & W_2(k) \\[2ex]
W_2'(k) & W_3(k)
\end{bmatrix}
=
\begin{bmatrix}
A_{11}' & A_{21}' \\[2ex]
A_{12}' & A_{22}'
\end{bmatrix}
\begin{bmatrix}
G_1(k+1) & G_2(k+1) \\[2ex]
G_2'(k+1) & G_3(G+1)
\end{bmatrix}
\begin{bmatrix}
A_{11} & A_{12} \\[2ex]
A_{21} & A_{22}
\end{bmatrix}
$$

$$\ldots \;(4.23)$$

Note that (4.22) is identical to (4.12) when $\bar{P}$ and $\bar{F}$ are replaced by P and F respectively. Hence (4.12) is referred to as the outer series equation. The equation for the boundary layer correction series is given by (4.23).

Now it is shown that the expressions for $\bar{F}(k+1)$ and $\bar{G}(k+1)$ can be expressed as power series in h (Appendix 4.3).

4.4 Development of the Method [34]

Once the equations for the outer series (4.22) and for the BLC series (4.23) are formulated, the method of developing the series solutions is straightforward. Firstly, the outer solution is assumed as a power series in h and the controlling equations are obtained. Secondly, a similar procedure is adopted for BLC solution. Finally, the total series solution and the terminal conditions for solving the series equations are obtained. The detailed procedure is given below.

Outer Series:

Let the outer series be represented as

$$\bar{P}(k) = \bar{P}_i^{(0)}(k) + h\,\bar{P}_i^{(1)}(k) + \ldots \ldots i = 1,2,3 \qquad \ldots \text{(4.24)}$$

Substituting (4.24) in (4.22) and collecting coefficients of like powers of h, the following series equations are obtained.

For zeroth order approximation

$$
\begin{bmatrix} \bar{P}_1^{(0)}(k) & P_2^{(0)}(k) \\ \bar{P}_2^{(0)\,'}(k) & \bar{P}_3^{(0)}(k) \end{bmatrix}
=
\begin{bmatrix} Q_1 & Q_2 \\ Q_2' & Q_3 \end{bmatrix}
$$

$$
+
\begin{bmatrix} A_{11}' & A_{21}' \\ A_{12}' & A_{22}' \end{bmatrix}
\begin{bmatrix} \bar{F}_1^{(0)}(k+1) & \bar{F}_1^{(0)}(k+1) \\ \bar{F}_2^{(0)\,'}(k+1) & \bar{F}_3^{(0)}(k+1) \end{bmatrix}
\begin{bmatrix} A_{11} & A_{12} \\ A_{21} & A_{22} \end{bmatrix}
\qquad \ldots \text{(4.25a)}
$$

For first order approximation

For first order approximation

$$
\begin{bmatrix} \bar{P}_1^{(1)}(k) & \bar{P}_2^{(1)}(k) \\[2ex] \bar{P}_2^{(1)'}(k) & \bar{P}_3^{(1)}(k) \end{bmatrix}
=
\begin{bmatrix} A'_{11} & A'_{21} \\[2ex] A'_{12} & A'_{22} \end{bmatrix}
\begin{bmatrix} \bar{F}_1^{(1)}(k+1) & \bar{F}_2^{(1)}(k+1) \\[2ex] \bar{F}_2^{(1)'}(k+1) & \bar{F}_3^{(1)}(k+1) \end{bmatrix}
\begin{bmatrix} A_{11} & A_{12} \\[2ex] A_{21} & A_{22} \end{bmatrix}
$$

$$\dots \quad (4.25b)$$

Similar equations are obtained for higher order approximations.

In (4.25), $\bar{F}^{(0)}(k+1)$ and $\bar{F}^{(1)}(k+1)$ represent the zeroth and first order coefficient matrices (Appendix 4.3).

Correction Series:

Let the correction series be represented as

$$W_i(k) = W_i^{(0)}(k) + hW_i^{(1)}(k) + \dots \dots \quad i = 1,2,3 \qquad \dots \quad (4.26)$$

Inserting (4.26) in (4.23) and collecting coefficients of like powers of h we get for zeroth order approximation

$$
\begin{bmatrix} 0 & W_2^{(0)}(k) \\[2ex] W_1^{(0)'}(k) & W_3^{(0)}(k) \end{bmatrix}
=
\begin{bmatrix} A'_{11} & A'_{21} \\[2ex] A'_{12} & A'_{22} \end{bmatrix}
\begin{bmatrix} G_1^{(0)}(k+1) & G_2^{(0)}(k+1) \\[2ex] G_2^{(0)'}(k+1) & G_3^{(0)}(k+1) \end{bmatrix}
\begin{bmatrix} A_{11} & A_{12} \\[2ex] A_{21} & A_{22} \end{bmatrix}
$$

$$\dots \quad (4.27a)$$

For first order approximation

$$
\begin{bmatrix} W_1^{(0)}(k) & W_2^{(1)}(k) \\[2ex] W_2^{(1)'}(k) & W_3^{(1)}(k) \end{bmatrix}
=
\begin{bmatrix} A'_{11} & A'_{21} \\[2ex] A'_{12} & A'_{21} \end{bmatrix}
\begin{bmatrix} G_1^{(1)}(k+1) & G_2^{(1)}(k+1) \\[2ex] G_2^{(1)'}(k+1) & G_3^{(1)}(k+1) \end{bmatrix}
\begin{bmatrix} A_{11} & A_{12} \\[2ex] A_{21} & A_{22} \end{bmatrix}
$$

$$\dots \quad (4.27b)$$

Higher order equations are obtained in a similar way.

In (4.27), $G^{(0)}(k+1)$ and $G^{(1)}(k+1)$ represent the zeroth and first order coefficient matrices (Appendix 4.3).

Total Series Solution:

The total series solution is obtained by proper summing of the outer series (4.24) and the correction series (4.26). That is

$$P_1(k) = (\bar{P}_1^{(0)}(k) + h\bar{P}_1^{(1)}(k) + \ldots \ldots) + h^{N-k+1}(W_1^{(0)}(k)+hW_1^{(1)}(k) + \ldots)$$

$$\ldots \quad (4.28a)$$

$$P_i(k) = (\bar{P}_i^{(0)}(k) + h\bar{P}_i^{(1)}(k) + \ldots \ldots)$$

$$+ h^{N-k}(W_i^{(0)}(k) + h\ W_i^{(1)}(k) + \ldots \ldots), \ i = 2,3$$

$$\ldots \quad (4.28b)$$

Final Conditions:

The final conditions required to solve the outer series equation (4.25) and the final BLC series equations (4.27) are obtained based on the fact that the total series solution (4.28) should satisfy the given final conditions. That is

$$h^0 : \quad \bar{P}_1^{(0)}(N) \ = \ P_1(N) \quad ;W^{(0)}(N)=P_i(N) \ - \ \bar{P}_i^{(0)}(N) \ , \ i = 2,3$$

$$\ldots \quad (4.29a)$$

$$h^1 : \quad \bar{P}_1^{(1)}(N) \ = \ - \ W_1^{(0)}(N) \quad ; \quad W_i^{(1)}(N) \ - \ \bar{P}_i^{(1)}(N), \ i = 2,3$$

$$\ldots \quad (4.29b)$$

Important Note:

Due to the association of coefficients h^{N-k+1} and h^{N-k} with the correction series in the total series solution (4.28), the correction series equation (4.27) need to be solved for only a few values of k near the final point utilizing the inverse of the reduced order matrix (for $G^{(j)}(k+1)$) which is already evaluated for the solution of the outer series equation (4.25). Thus the solution of the correction series equation (4.27) does not require any extra matrix inversions and is obtained by simple recursion. This aspect of the correction series offers a considerable reductiion in the overall computation for obtaining the approximate solutions.

Example 4.2

Consider a second order discrete control system

$$
\begin{bmatrix} y(k+1) \\ z(k+1) \end{bmatrix} = \begin{bmatrix} 0.95 & h \\ 1.0 & 0 \end{bmatrix} \begin{bmatrix} y(k) \\ z(k) \end{bmatrix} + \begin{bmatrix} 1 \\ 0 \end{bmatrix} u(k) \qquad \ldots \quad (4.30)
$$

where the small parameter h = 0.12
The initial conditions are

$$
y(0) = 10.0 \quad ; \quad z(0) = 15.0
$$

and the performance index is

$$
J = \frac{1}{2} \sum_{k=0}^{N-1} [x'(k) \, Qx(k) + R \, u(k)^2]
$$

where $Q = \begin{bmatrix} 1 & 0 \\ 0 & 2h^2 \end{bmatrix}$; $x(k) = \begin{matrix} y(k) \\ z(k) \end{matrix}$; R = 2 and N = 8

$$
BR^{-1}B' = \begin{bmatrix} 0.5 & 0 \\ 0 & 0 \end{bmatrix}, \text{ where } B = \begin{bmatrix} 1 \\ 0 \end{bmatrix}
$$

The singularly perturbed matrix Riccati difference equation corresponding to (4.10) becomes

$$
\begin{bmatrix} p_1(k) & hp_2(k) \\ hp_2'(k) & h^2 p_3(k) \end{bmatrix} = \begin{bmatrix} 1 & 0 \\ 0 & 2h^2 \end{bmatrix} + \begin{bmatrix} 0.95 & 1.0 \\ h & 0 \end{bmatrix}
$$

$$
\left[\left\{ \begin{matrix} p_1(k+1) & hp_2(k+1) \\ hp_2'(k+1) & h^2 p_3(k+1) \end{matrix} \right\}^{-1} + \left\{ \begin{matrix} 0.5 & 0 \\ 0 & 0 \end{matrix} \right\} \right]^{-1} \begin{bmatrix} 0.95 & h \\ 1 & 0 \end{bmatrix}
$$

$$
\ldots \quad (4.31)
$$

The closed-loop optimal control corresponding to (4.9) is

$$u(k) = -0.5[1 \quad 0] \left[\left\{ \begin{pmatrix} p_1(k+1) & hp_2(k+1) \\ hp_2'(k+1) & h^2p_3(k+1) \end{pmatrix} \right\}^{-1} \right.$$

$$\left. + \left\{ \begin{pmatrix} 0.5 & 0 \\ 0 & 0 \end{pmatrix} \right\} \right]^{-1} \begin{bmatrix} 0.95 & h \\ 1 & 0 \end{bmatrix} \begin{bmatrix} y(k) \\ z(k) \end{bmatrix}$$

$$\dots \quad (4.32)$$

Proceeding on the lines suggested above, the series equations for zeroth and first order approximations are obtained as shown below.

Outer Series:

For zeroth order

$$p_1^{(0)}(k) = \frac{2 + 2.805 \ p_1^{(0)}(k+1)}{2 + p_1^{(0)}(k+1)} \qquad \dots \quad (4.33a)$$

$$p_2^{(0)}(k) = \frac{1.9 \ p_1^{(0)}(k+1)}{2 + p_1^{(0)}(k+1)} \qquad \dots \quad (4.33b)$$

$$p_3^{(0)}(k) = \frac{4 + 4p_1^{(0)}(k+1)}{2 + p_1^{(0)}(k+1)} \qquad \dots \quad (4.33c)$$

For first order

$$p_1^{(1)}(k) = \frac{p_1^{(1)}(k+1)(2.805 - p_1^{(0)}(k)) + 3.8p_2^{(0)}(k+1)}{2 + p_1^{(0)}(k+1)} \qquad \dots \quad (4.34a)$$

$$p_2^{(1)}(k) = \frac{p_1^{(1)}(k+1)(1.9 - p_2^{(0)}(k)) + 2p_2^{(0)}(k+1)}{2 + p_1^{(0)}(k+1)} \qquad \dots \quad (4.34b)$$

$$p_3^{(1)}(k) = \frac{p_1^{(1)}(k+1)(4 - p_3^{(0)}(k))}{2 + p_1^{(0)}(k+1)} \qquad \dots \quad (4.34c)$$

Correction Series:

For zeroth order

$$0 = w_1^{(0)}(k+1)(2.805 - p_1^{(0)}(k))+3.8w_2^{(0)}(k+1)$$

... (4.35a)

$$w_2^{(0)}(k) = \frac{w_1^{(0)}(k+1)(1.9-p_2^{(0)}(k)-h^{N-k}w_2^{(0)}(k))+2w_2^{(0)}(k+1)}{2 + p_1^{(0)}(k+1)}$$

... (4.35b)

$$w_3^{(0)}(k) = \frac{w_1^{(0)}(k+1)(4-h^{N-k}w_3^{(0)}(k))-w_1^{(0)}(k)p_3^{(0)}(k)}{2 + p_1^{(0)}(k+1)}$$

... (4.35c)

The total series solution is given by

$$p_1(k) = p_1^{(0)}(k)+hp_1^{(1)}(k)+h^2p_1^{(2)}(k) + \ldots\ldots$$

$$+ h^{N-k+1} \quad w_1^{(0)}(k)+hw_1^{(1)}(k) + \ldots\ldots$$

... (4.36a)

$$p_i(k) = p_i^{(0)}(k) + hp_i^{(1)}(k) + h^2p_i^{(2)}(k) + \ldots\ldots$$

$$+ h^{N-k} [w_i^{(0)}(k) + hw_i^{(1)}(k) + \ldots\ldots], \, i = 2,3$$

Note: In (4.35), the terms associated with h^{N-k} are to be handled carefully as eplained in section 4.1. It is evident that the correction series equations are to be solved for k less than N only. Therefore those terms associated with h^{N-k} should not be considered for zeroth order approximation.

The following simple algorithm will indicate the various steps involved in solving the series equations (4.34) and (4.35).

Algorithm

    (i) Choosing $p_1^{(0)}(8) = p_1(8) = 0$, solve (4.33a) and find $p_1^{(0)}(k)$. The solutions of $p_2^{(0)}(k)$ and $p_3^{(0)}(k)$ are fixed from (4.33b) and (4.33c) respectively.

(ii)   Choosing $w_2^{(0)}(8) = p_2(8) - p_2^{(0)}(8) = 0 - p_2^{(0)}(8)$, and

$\qquad\qquad w_3^{(0)}(8) = p_3(8) - p_3^{(0)}(8) = 0 - p_3^{(0)}(8)$

find $w_1^{(0)}(8)$ from (4.35a), and $w_2^{(0)}(7)$ and $w_3^{(0)}(7)$

from (4.35b) and (4.35c) respectively.

(iii)   Choosing $p_1^{(1)}(8) = - w_1^{(0)}(8)$, solve (4.34a) and find $p_1^{(0)}(k)$.

The solution $p_2^{(1)}(k)$ and $p_3^{(1)}(k)$ are fixed from (4.34b) and (4.34c) respectively.

The above computations are sufficient to get the total series solution upto first order approximation. The procedure is similar for second and higher order approximations.

It is <u>noted</u> from (4.36) that for first order total series solution, the first order correction series is not at all required. The low-order inverse determined for zeroth order solution is sufficient and no new inverses are needed either for the higher order outer series or for the correction series.

Using the algorithm, the Riccati coefficients p(k) corresponding to degenerate, zeroth and first order solutions are evaluated. The results are compared with the exact solution obtained from (4.31) in Table 4.2. Using the series solutions obtained, the closed loop optimal control and the state trajectories are evaluated from (4.32) and (4.30) respectively. The results are compared with the exact solution in Table 4.3. The performance indices are shown in Table 4.4.

TABLE - 4.2

Comparison of the approximate and exact solution of the
Matrix Riccati Equation of Example 4.2

p(k)	Degenerate solution	Zeroth order solution	First order solution	Exact Solution
$p_1(0)$	1.8728	1.8728	2.0151	2.0734
$p_2(0)$	0.9192	0.9192	1.0117	1.0275
$p_3(0)$	2.9653	2.953	3.0069	3.0208
$p_1(1)$	1.8726	1.8726	2.0146	2.0724
$p_2(1)$	0.9183	0.9183	1.0108	1.0267
$p_3(1)$	2.9653	2.9653	3.0069	3.0278
$p_1(2)$	1.8719	1.8719	2.0125	2.0695
$p_2(2)$	0.9175	0.9175	1.0092	1.0242
$p_3(2)$	2.9653	2.9653	3.0000	3.0139

TABLE - 4.2 Contd.

p(k)	Degenerate solution	Zeroth order solution	First order solution	Exact Solution
$p_1(3)$	1.8687	1.8687	2.0005	2.0601
$p_2(3)$	0.9142	0.9142	1.0008	1.0158
$p_3(3)$	2.9653	2.9653	2.9931	3.0069
$p_1(4)$	1.8557	1.8557	1.9770	2.0285
$p_2(4)$	0.9008	0.9008	0.9750	0.9867
$p_3(4)$	2.9514	2.9514	2.9722	2.9792
$p_1(5)$	1.8027	1.8027	1.8829	1.9270
$p_2(5)$	0.8450	0.8450	0.8875	0.8950
$p_3(5)$	2.8889	2.8889	2.8889	2.8958
$p_1(6)$	1.6017	1.6017	1.6017	1.6305
$p_2(6)$	0.6333	0.6333	0.6333	0.6333
$p_3(6)$	2.6667	2.6667	2.6667	2.6667
$p_1(7)$	1.0000	1.0000	1.0000	1.0000
$p_2(7)$	0.0000	0.0000	0.0000	0.0000
$p_3(7)$	2.0000	2.0000	2.0000	2.0000
$p_1(8)$	0.0000	0.0000	0.0000	0.0000
$p_2(8)$	-1.0525	0.0000	0.0000	0.0000
$p_3(8)$	0.8889	0.0000	0.0000	0.0000

TABLE 4.3

Comparison of approximate and exact solution of states and
optimal control of Example 4.2

y(k)/z(k) /u(k)	Degenerate solution	Zeroth order slution	First order solution	Exact solution
y(0)	10.0000	10.0000	10.0000	10.0000
z(0)	15.0000	15.0000	15.0000	15.0000
u(0)	-5.7488	-5.7488	-5.9727	-6.0529
y(1)	5.5512	5.5512	5.3273	5.2471
z(1)	10.0000	10.0000	10.0000	10.0000
u(1)	-3.2876	-3.2876	-3.3009	-3.3037
y(2)	3.1860	3.1860	2.9600	2.8811
z(2)	5.5512	5.5512	5.3273	5.2471
u(2)	-1.8741	-1.8741	-1.8147	-1.7947
y(3)	1.8187	1.8187	1.6365	1.5720
z(3)	3.1860	3.1860	2.9600	2.8811
u(3)	-1.0665	-1.0665	-0.9976	-0.9723
y(4)	1.0436	1.0436	0.9123	0.8668
z(4)	1.8187	1.8187	1.6365	1.5720
u(4)	-0.6013	-0.6013	-0.5405	-0.5204
y(5)	0.6084	0.6084	0.5226	0.4918
z(5)	1.0436	1.0436	0.9123	0.8668
u(5)	-0.3255	-0.3255	-0.2805	-0.2668

TABLE 4.3 Contd.

y(k)/z(k) /u(k)	Degenerate solution	Zeroth order slution	First order solution	Exact solution
y(6)	0.3776	0.3776	0.3254	0.3044
z(6)	0.6084	0.6084	0.5226	0.4918
z(6)	-0.1439	-0.1439	-0.1240	-0.1161
y(7)	0.2878	0.2878	0.2479	0.2321
y(7)	0.3776	0.3776	0.3254	0.3044
u(7)	0.0182	0.0000	0.0000	0.0000
y(8)	0.3214	0.3188	0.2746	0.2570
z(8)	0.2778	0.2878	0.2479	0.2321

TABLE - 4.4

Comparison of performance indices of approximate and exact solutions of Example 4.2

Nature of the solution	Performance index
Degenerate solution	127.3160
Zeroth order solution	127.3153
First order solution	127.0612
Exact solution	127.0424

Example 4.3 [24,36]

Consider the fourth order discrete control problem

$$
\begin{bmatrix} y_1(k+1) \\ y_2(k+1) \\ z_1(k+1) \\ z_2(k+1) \end{bmatrix} = \begin{bmatrix} 1.0 & 1.0 & 1.0h & 1.0h \\ 0.0 & 1.0 & 0.0h & 0.1h \\ 1.0 & -1.0 & 0.1h & 0.5h \\ 0.2 & 0.0 & 0.0h & 1.0h \end{bmatrix} \begin{bmatrix} y_1(k) \\ y_2(k) \\ z_1(k) \\ z_2(k) \end{bmatrix} + \begin{bmatrix} 1 \\ 1 \\ 0 \\ 0 \end{bmatrix} u(k)
$$

$$\dots \quad (4.37)$$

where the small parameter h = 0.1
The initial conditions are

$$
\begin{bmatrix} y_1(0) \\ y_2(0) \end{bmatrix} = \begin{bmatrix} 10.0 \\ 8.0 \end{bmatrix} \quad ; \quad \begin{bmatrix} z_1(0) \\ z_2(0) \end{bmatrix} = \begin{bmatrix} 5.0 \\ 2.0 \end{bmatrix}
$$

and the performance index is

$$
j = \frac{1}{2} \sum_{k=0}^{N-1} [x'(k) \ Qx(k) + u'(k) \ Ru(k)]
$$

where

$$
Q = \begin{bmatrix} 0 & 0 & 0 & 0 \\ 0 & 1 & 0 & 0 \\ 0 & 0 & 0 & 0 \\ 0 & 0 & 0 & 0 \end{bmatrix} \quad ; \quad x(k) = \begin{bmatrix} y_1(k) \\ y_2(k) \\ z_1(k) \\ z_2(k) \end{bmatrix}
$$

$$
R = 1.0 \quad \text{and} \quad N = 12
$$

The eigenvalues of (4.12) are

$p_1 = 1.1238$; $p_2 = 0.98549$; $p_3 = -0.09298$; $p_4 = 0.084659$

Following the method developed, the degenerate, zeroth and first order series equations of the matrix Riccati equation are solved and the series solutions obtained are compared with the exact solution in Fig. 4.1. Using the series solutions obtained as above, the optimal control and the state trajectories are found from (4.9) and (4.7) respectively. The optimal control of series solutions are compared with the exact optimal control in Table 4.5. The state trajectories are shown in Fig 4.2. Finally, the performance indices are evaluated for degenerate, zeroth, and first order solutions and compared with the exact performance index in Table 4.6.

With the increased order of approximation, the series solutions approach the exact solution very closely.

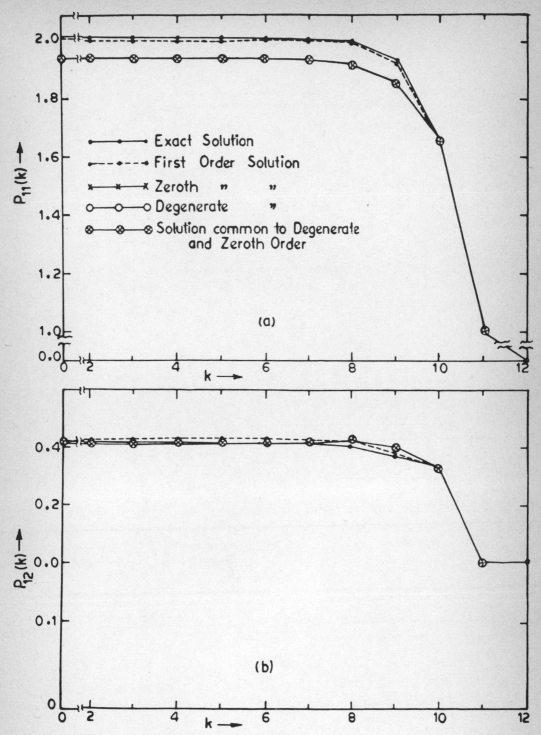

Fig. 4.1(a) & (b)   Exact and approximate solutions of $p_{11}(k)$ & $p_{12}(k)$
of Example 4.3

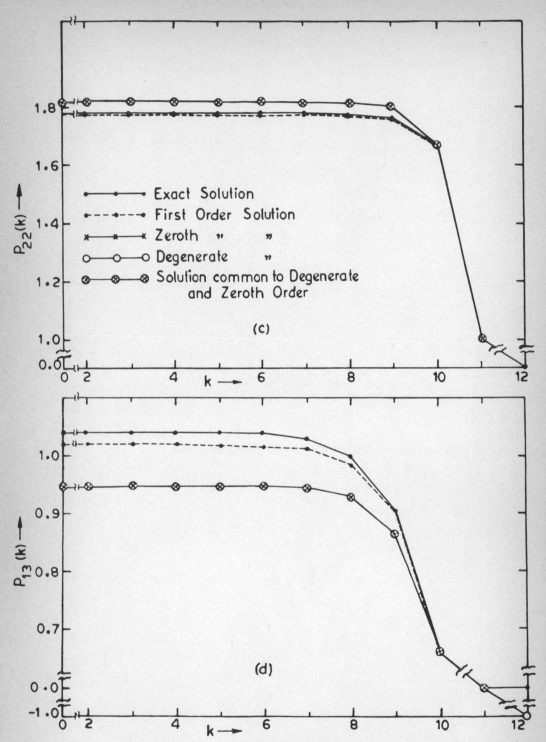

Fig. 4.1(c) & (d)   Exact and approximate solutions of $p_{22}(k)$ & $p_{13}(k)$
of Example 4.3

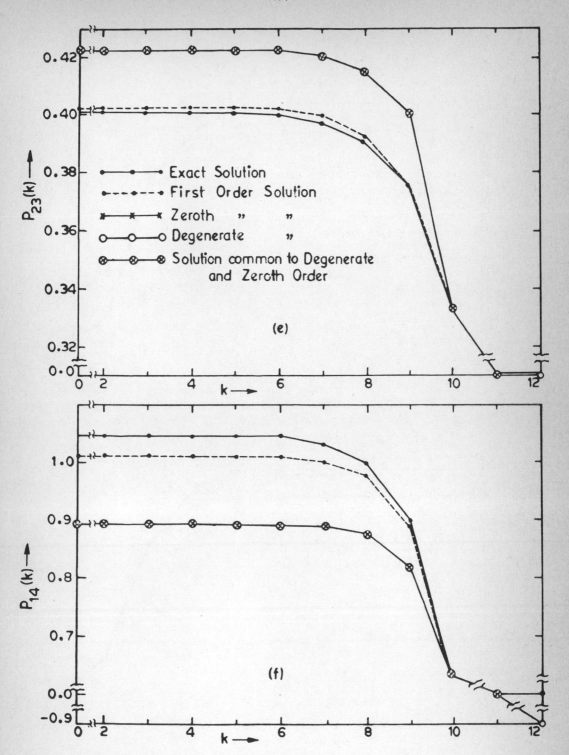

Fig. 4.1(e) & (f)   Exact and approximate solutions of $p_{23}(k)$ & $p_{14}(k)$ of Example 4.3

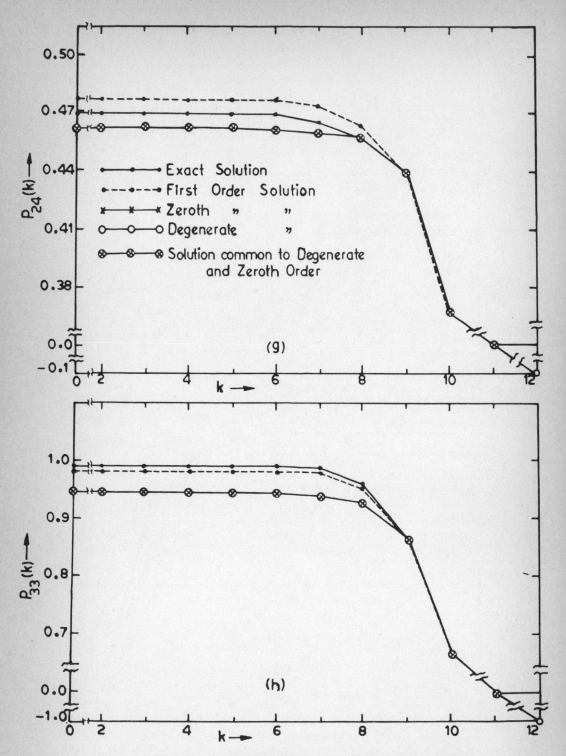

Fig. 4.1(g) & (h)   Exact and approximate solutions of $p_{24}(k)$ & $p_{33}(k)$
of Example 4.3

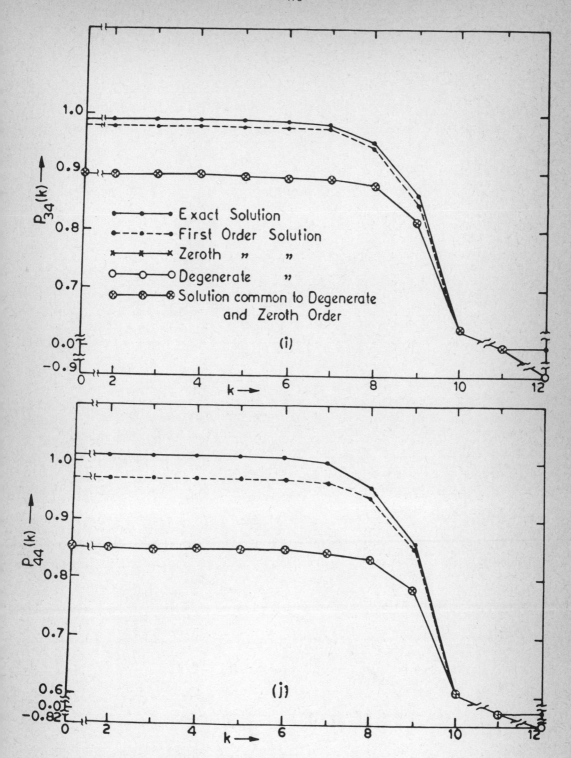

Fig. 4.1(i) & (j) Exact and approximate solutions of $P_{34}(k)$ & $P_{44}(k)$
of Example 4.3

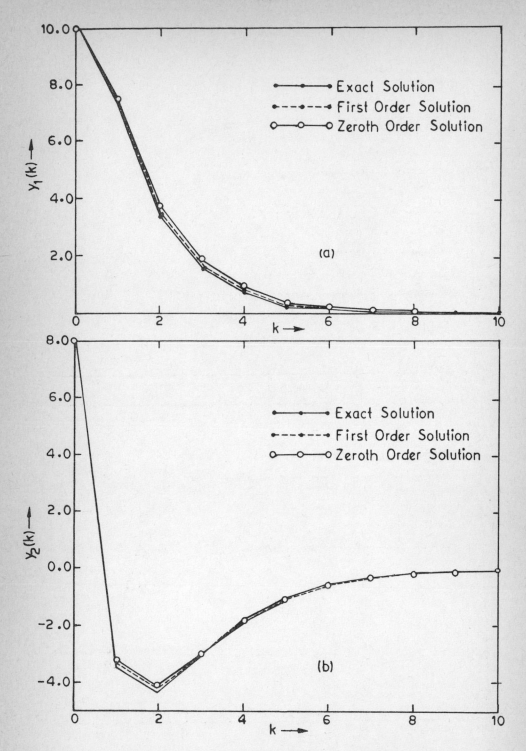

Fig. 4.2(a) & (b)   Exact and approximate solutions of $y_1(k)$ & $y_2(k)$ of
Example 4.3

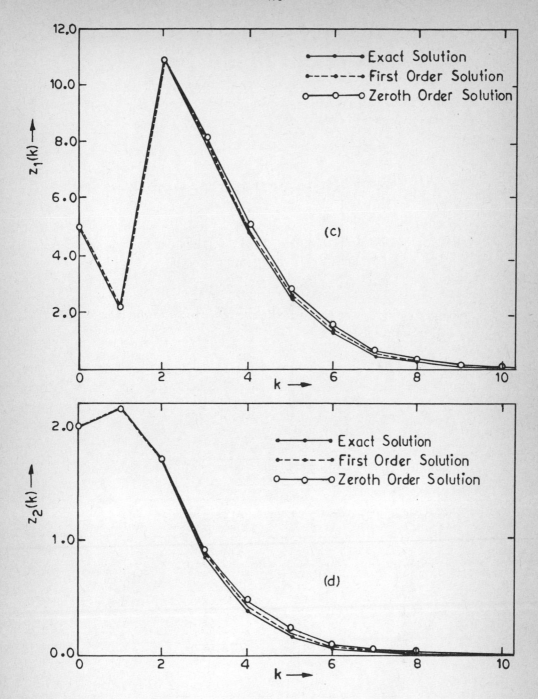

Fig. 4.2(c) & (d)   Exact and approximate solutions of $z_1(k)$ & $z_2(k)$ of Example 4.3

4.5 Steady State Solution

In many practical situations, the steady state solution of the matrix Riccati equation (4.10) is of vital importance [41]. The steady state equation is an algebraic equation which is obtained from (4.12) by writing $P(k+1) = P(k) = P$.

That is

$$
\begin{bmatrix} P_1 & P_2 \\ \\ P'_2 & P_3 \end{bmatrix} = \begin{bmatrix} Q_1 & Q_2 \\ \\ Q'_2 & Q_3 \end{bmatrix} + \begin{bmatrix} A'_{11} & A'_{21} \\ \\ A'_{12} & A'_{22} \end{bmatrix} \begin{bmatrix} F_1 & hF_2 \\ \\ hF'_2 & h^2F_3 \end{bmatrix} \begin{bmatrix} A_{11} & A_{12} \\ \\ A_{21} & A_{22} \end{bmatrix}
$$

$$\dots \quad (4.38)$$

In order to develop a singular perturbation method for (4.38), we write the outer series as

$$
P_i = P_i^{(0)} + h\, P_i^{(0)} + \dots \dots , \quad i = 1,2,3 \qquad \dots \quad (4.39)
$$

TABLE - 4.5

Comparison of approximate and exact optimal control of Example 4.3

u(k)	Degenerate solution	Zeroth order solution	First order solution	Exact solution
u(0)	-11.2047	-11.2047	-11.3912	-11.4260
u(1)	- 1.0440	- 1.0440	- 0.9284	- 0.9049
u(2)	1.1394	1.1394	1.2481	1.2680
u(3)	1.1911	1.1911	1.2386	1.2463
u(4)	0.8172	0.8172	0.8203	0.8199
u(5)	0.4837	0.4837	0.4672	0.4637
u(6)	0.2651	0.2651	0.2447	0.2409
u(7)	0.1386	0.1386	0.1213	0.1182
u(8)	0.0701	0.0701	0.0576	0.0555
u(9)	0.0341	0.0341	0.0262	0.0248
u(10)	0.0147	0.0147	0.0102	0.0095
u(11)	0.0209	0.0000	0.0000	0.0000

TABLE 4.6

Comparison of performance indices of series and exact solutions of Example 4.3

Degeneration solution	-	205.46780
Zeroth order solution	-	205.46736
First order solution	-	205.24781
Exact solution	-	205.21266

Insertion of (4.39) in (4.38) and collection of coefficients of like powers of h on either side, we get for zeroth order approximation

$$\begin{bmatrix} P_1^{(0)} & P_2^{(0)} \\ P_2^{(0)} & P_3^{(0)} \end{bmatrix} = \begin{bmatrix} Q_1 & Q_2 \\ Q_2' & Q_3 \end{bmatrix} + \begin{bmatrix} A_{11}' & A_{21}' \\ A_{12}' & A_{22}' \end{bmatrix} \begin{bmatrix} F_1^{(0)} & 0 \\ 0 & 0 \end{bmatrix} \begin{bmatrix} A_{11} & A_{12} \\ A_{21} & A_{22} \end{bmatrix}$$

$$\dots \quad (4.40)$$

where $F_1^{(0)} = P_1^{(0)} [I_1 + E_1 P_1^{(0)}]^{-1}$

Rearranging (4.32a), we get

$$\left. \begin{aligned} P_1^{(0)} &= Q_1 + A'_{11} F_1^{(0)} A_{11} \\ P_2^{(0)} &= Q_2 + A'_{11} F_1^{(0)} A_{12} \\ P_3^{(0)} &= Q_3 + A'_{12} F_1^{(0)} A_{12} \end{aligned} \right\} \qquad \dots \quad (4.41)$$

It is clear from (4.41) that the first one is an algebraic equation in $P_1^{(0)}$ of order n_1 and decoupled from $P_2^{(0)}$ and $P_3^{(0)}$ which are obtained once $P_1^{(0)}$ is known. On the otherhand the original steady state equation (4.38) is an algebraic equation of order (n_1+n_2) and coupled w.r.t. P_1, P_2, and P_3.

For first order approximation

$$\begin{bmatrix} P_1^{(1)} & P_2^{(1)} \\ P_2^{(1)'} & P_3^{(1)} \end{bmatrix} = \begin{bmatrix} A_{11}' & A_{21}' \\ A_{12}' & A_{22}' \end{bmatrix} \begin{bmatrix} F_1^{(1)} & F_2^{(0)} \\ F_2^{(0)'} & 0 \end{bmatrix} \begin{bmatrix} A_{11} & A_{12} \\ A_{21} & A_{22} \end{bmatrix}$$

$$\dots \quad (4.42)$$

Similar equations are obtained for higher order approximations.

Once again (4.42) is an algebraic equation in $P_1^{(1)}$ and $P_2^{(1)}$ and $P_3^{(1)}$ are obtained once $P_1^{(1)}$ is known.

In the steady state case, there is no finite final condition specified for P and hence, there is no need for any final boundary layer correction series. The total series solution is given by the outer series itself (4.39).

Example 4.4

Consider the <u>Steady state</u> case of Example 4.3. The matrix Riccati algebraic equation is solved for zeroth and first order approximations. The series solutions and the exact steady state solution are obtained as furnished below.

The zeroth order solution of the Riccati matrix is

$$
\begin{bmatrix}
1.94712 & 0.42208 & 0.09471 & 0.08946 \\
0.42208 & 1.82185 & 0.04221 & 0.04621 \\
0.09471 & 0.04221 & 0.00947 & 0.00895 \\
0.08946 & 0.04621 & 0.00895 & 0.00851
\end{bmatrix}
$$

The first order solution is

$$
\begin{bmatrix}
2.05944 & 0.42964 & 0.10198 & 0.10118 \\
0.42964 & 1.77564 & 0.04028 & 0.04767 \\
0.10198 & 0.04028 & 0.00980 & 0.00973 \\
0.10118 & 0.04767 & 0.00973 & 0.00973
\end{bmatrix}
$$

The exact solution is

$$
\begin{bmatrix}
2.09185 & 0.42208 & 0.10394 & 0.10477 \\
0.42208 & 1.78101 & 0.04010 & 0.04690 \\
0.10394 & 0.04010 & 0.00995 & 0.00995 \\
0.10477 & 0.04690 & 0.00995 & 0.01013
\end{bmatrix}
$$

Using the steady state values of Riccati coefficients shown above, the state trajectories and the optimal control are evaluated for zeroth and first order approximations. The results are compared with the exact steady state solutions in Table 4.7 (optimal control) and Fig. 4.3 (state trajectories). The performance indices are shown in Table 4.8.

4.6 Conclusions

In this chapter, firstly, a method has been proposed to analyze the singularly perturbed nonlinear difference equations for initial and boundary value problems. The approximate solutions have been obtained in terms of the outer series and the correction series. It has been found that considerable care has to be taken in the formulation of the equations controlling the boundary layer correction series.

Secondly, the closed-loop optimal control problem for a linear shift invariant discrete system has been formulated. The corresponding matrix Riccati difference equation is cast in the singularly perturbed form. The approximate solutions have been obtained by a method, requiring the outer series and the final BLC series to take care of the lost final conditions in the process of degeneration [34].

Finally, a method has been developed for the more important case of the steady state matrix Riccati algebraic equation.
Numerical example have been provided to illustrate the proposed methods.

The special features of this chapter in developing singular perturbation methods for the closed-loop optimal control are
(i) In solving the high order matrix Riccati difference equation (4.12), order reduction has been effectively used in evaluating the inverse of the matrix in F of order (n_1+n_2) given by (4.13). That is, the solution of the zeroth order outer series equation (4.25a) requires the inverse of the matrix, i.e., $[I_1+E_1 P_1^{(0)}(k+1)]^{-1}$, which is of order n_1 only. In addition, the first order equation (4.25b) (or high order equation) uses the same reduced order matrix inversion and hence is solved by simple recursion.

This is also true for the steady state case of the matrix Riccati equation.
(ii) The correction series equation (4.27) need to be solved for only a few values of k near the final boundary layer, using the reduced order matrix inversion which is already evaluated for the solution of

180

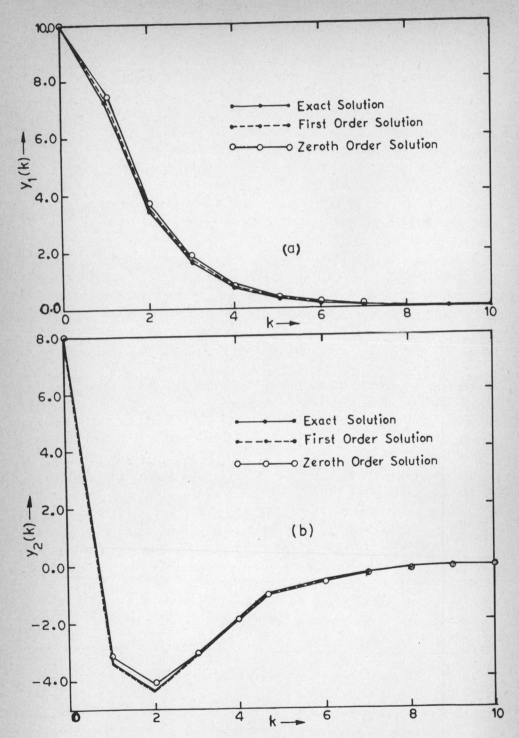

Fig. 4.3(a) & (b)   Exact and approximate solutions of $y_1(k)$ & $y_2(k)$
of Example 4.4

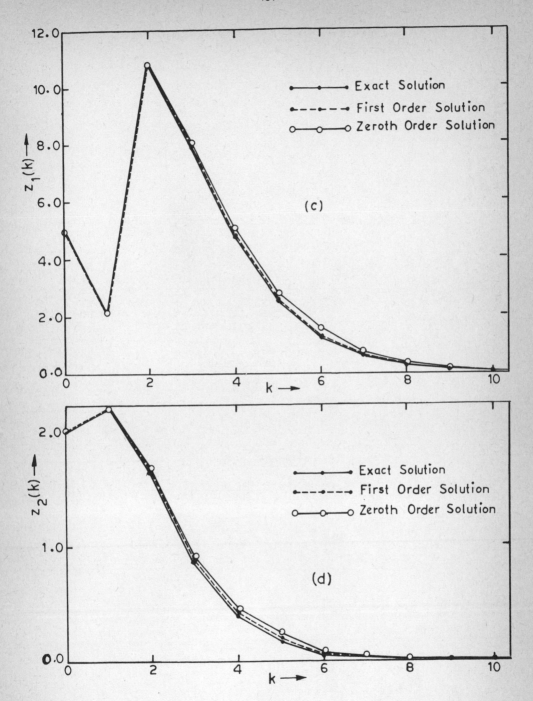

Fig. 4.3(c) & (d)   Exact and approximate solutions of $z_1(k)$ & $z_2(k)$ of Example 4.4

the outer series equation (4.25). As such, the solution of correction series does not demand any more extra matrix inversions.

These two features are mainly responsible for weighing in favour of the proposed methods which offer considerable reduction in the overall compution.

TABLE - 4.7

Comparison of approximate and exact optimal control of
Example 4.4

u(k)	Zeroth Order solution	First order solution	Exact solution
u(0)	-11.2047	-11.3912	-11.4260
u(1)	- 1.0440	- 0.9284	- 0.9049
u(2)	1.1394	1.2481	1.2680
u(3)	1.1911	1.2386	1.2463
u(4)	0.8172	0.8203	0.8199
u(5)	0.4837	0.4672	0.4636
u(6)	0.2651	0.2447	0.2409
u(7)	0.1385	0.1212	0.1182
u(8)	0.0700	0.0576	0.0555
u(9)	0.0345	0.0264	0.0251
u(10)	0.0167	0.0118	0.0110
u(11)	0.0080	0.0051	0.0047

TABLE - 4.8

Comparison of performance indices of
series and exact solutions of Example 4.4

Zeroth order solution	-	205.5675
First order slution	-	205.2479
Exact solution	-	205.2428

APPENDIX 4.1

Using (4.11) in (4.10)

$$
\begin{bmatrix} P_1(k) & hP_2(k) \\ hP_2'(k) & h^2P_3(k) \end{bmatrix} = \begin{bmatrix} Q_1 & hQ_2 \\ hQ_2' & h^2Q_3 \end{bmatrix} +
$$

$$
\begin{bmatrix} A_{11}' & A_{21}' \\ hA_{12}' & hA_{22}' \end{bmatrix} \begin{bmatrix} F_1(k+1) & F_2(k+1) \\ F_2'(k+1) & F_3(k+1) \end{bmatrix} \begin{bmatrix} A_{11} & hA_{12} \\ A_{21} & hA_{22} \end{bmatrix} \quad \ldots \quad (A4.1)
$$

where the matrix $F(k+1)$ is given by (4.13). Note that F is symmetric, since $BR^{-1}B'$ and P are symmetric.
Rearranging (A4.1)

$$
P_1(k) = Q_1 + A_{11}'(F_1(k+1) A_{11} + F_2(k+1)A_{21})
$$
$$
+ A_{21}'(F_2'(k+1) A_{11} + F_3(k+1)A_{21}) \quad \ldots \quad (A4.2a)
$$

$$
hP_2(k) = hQ_2 + hA_{11}'(F_1'(k+1)A_{12} + F_2(k+1)A_{22})
$$
$$
+ hA_{21}'(F_2'(k+1)A_{12} + F_3(k+1)A_{22}) \quad \ldots \quad (A4.2b)
$$

$$
h^2P_3(k) = h^2Q_3 + h^2A_{12}'(F_1(k+1)A_{12} + F_2(k+1)A_{22})
$$
$$
+ h^2 A_{22}'(F_2'(k+1)A_{12} + F_3(k+1)A_{22}) \quad \ldots \quad (A4.2c)
$$

Dividing by h and h^2 throughout the equations (A4.2b) and (A4.2c) respectively, the resulting equation becomes (4.12). From (A4.2b) and (A4.2c) it is now clear why the matrices, P, Q, and S are partitioned as shown in (4.13).

APPENDIX 4.2

In (4.20a), consider

$$[I + E\bar{P} + h^{N-k}EW]^{-1} = [\ (I+E\bar{P})\ (I + h^{N-k}\ (I+E\bar{P})^{-1}EW)\ ]^{-1}$$

$$= [\ I+h^{N-k}(I+E\bar{P})^{-1}\ EW]^{-1}\ [I+E\bar{P}]^{-1}$$

$$= [\ I+h^{N-k}\ CEW]^{-1}C,\ C = (I+E\bar{P})^{-1} \quad \ldots \text{(A4.3)}$$

When $k<N$, $h^{N-k} \ll 1$, h being the singular perturbation parameter. Therefore the norm of $h^{N-k}CEW$ is much less than unity. Ignoring terms with h and higher powers of h, (A4.3) becomes

$$[I+E\bar{P} + h^{N-k}EW]^{-1} = (I - h^{N-k}CEW)C$$

$$= C - h^{N-k}CEWC \quad \ldots \text{(A4.4)}$$

Substituting (A4.4) in (4.20a),

$$F = (\bar{P} + h^{N-k}W)\ (C - h^{N-k}\ CEWC)$$

$$= \bar{P}\ C + h^{N-k}\ (WC - PCEWC - h^{N-k}\ WCEWC)$$

$$= \bar{F} + h^{N-k}\ G \quad \ldots \text{(A4.5)}$$

(A4.5) is the same as (4.20b).

APPENDIX 4.3

The matrix $\bar{F}(k+1)$ can be expressed as a power series in h as shown below. Consider the expression for $F(k+1)$ in (4.20b), (The arguments k and (k+1) are omitted for the sake of notational simplicity)

$$\bar{F} = \bar{P}\ (I + E\bar{P})^{-1} \quad \ldots \text{(A4.6)}$$

Consider

$$(I + E\bar{P})^{-1} = \left[\left\{ \begin{array}{cc} I_1 & 0 \\ 0 & I_2 \end{array} \right\} + \left\{ \begin{array}{cc} E_1 & E_2 \\ E_2' & E_3 \end{array} \right\} \left\{ \begin{array}{cc} \bar{P}_1 & h\bar{P}_2 \\ h\bar{P}_2' & h^2\bar{P}_3 \end{array} \right\} \right]^{-1}$$

$$= \left[\underbrace{\left\{ \begin{array}{cc} I_1 + E_1\bar{P}_1 & 0 \\ E_2'\bar{P}_1 & I_2 \end{array} \right\}}_{L} + h\underbrace{\left\{ \begin{array}{cc} E_2\bar{P}_2' & E_1\bar{P}_2 \\ E_3\bar{P}_2' & E_2'\bar{P}_2 \end{array} \right\}}_{M} + h^2\underbrace{\left\{ \begin{array}{cc} 0 & E_2\bar{P}_3 \\ 0 & E_3\bar{P}_3 \end{array} \right\}}_{N} \right]^{-1}$$

$$= (L + hM + h^2N)^{-1}$$

$$= [L(I + hL^{-1}M + h^2L^{-1}N)]^{-1}$$

$$= (I + hL^{-1}M + h^2L^{-1}N)^{-1} L^{-1} \qquad \ldots \text{(A4.7)}$$

For sufficiently small values of h, the norms of $hL^{-1}M$ and $h^2L^{-1}N$ are much less than unity. Therefore (A4.7) becomes

$$(I+E\bar{P})^{-1} = L^{-1} - h\,L^{-1}ML^{-1} + 0\,(h^2) \qquad \ldots \text{(A4.8)}$$

Since L is a triangular matrix, its inverse is readily written as [45],

$$L^{-1} = \left[\begin{array}{cc} I_1 + E_1\bar{P}_1 & 0 \\ E_2'\,\bar{P}_1 & I_2 \end{array} \right]^{-1} = \left[\begin{array}{cc} (I_1 + E_1\bar{P}_1)^{-1} & 0 \\ -E_2'\,\bar{P}_1(I_1+E_1\bar{P}_1)^{-1} & I_2 \end{array} \right]$$

$$\ldots \text{(A4.9)}$$

Using (A4.8) and (A4.9) in (A4.6),

$$\bar{F} = \left[\begin{array}{cc} \bar{F}_1 & \bar{F}_2 \\ \bar{F}_2' & \bar{F}_3 \end{array} \right] = \left[\begin{array}{cc} \bar{P}_1 & h\bar{P}_2 \\ h\bar{P}_2' & h^2\bar{P}_3 \end{array} \right] \left\{ \left\{ \begin{array}{cc} C_1 & 0 \\ -E_2'\,\bar{P}_1C_1 & 0 \end{array} \right\} - h \left[\begin{array}{cc} C_1 & 0 \\ -E_2'\bar{P}_1C_1 & I_2 \end{array} \right] \right.$$

$$\left. \left\{ \begin{array}{cc} E_2\,\bar{P}_2'\,C_1 - E_1\,\bar{P}_2\,E_2'\,\bar{P}_1\,C_1 & E_1'\,\bar{P}_2 \\ E_3\,\bar{P}_2\,C_1 - E_2'\,P_2\,E_2'\,\bar{P}_1\,C_1 & E_2'\,\bar{P}_2 \end{array} \right\} \right]$$

$$\ldots \text{(A4.10)}$$

where $C_1 = (I_1 + E_1 \bar{P}_1)^{-1}$

Expanding (A4.10),

$$\bar{F}_1 = \bar{P}_1 C_1 - h\bar{P}_1 C_1(E_2\bar{P}_2' C_1 - E_1\bar{P}_2E_2' \bar{P}_1C_1) - h \bar{P}_2 E_2'\bar{P}_1C_1$$

$$+ h(I_1 - \bar{P}_1 C_1 E_1) \bar{P}_2 + 0 (h^2)$$

$$\bar{F}_2 = h(I_1 - \bar{P}_1 C_1 E_1) \bar{P}_2 + 0(h^2)$$

$$\bar{F}_2' = h \bar{P}_2' C_1 + 0(h^2)$$

$$\bar{F}_3 = 0(h^2) \qquad \qquad \cdots \ (A4.11)$$

For the matrix $\bar{F}$ to be symmetric, it is necessary to show that from (A.411),

$$(\bar{P}_2' C_1)' = (I_1 - \bar{P}_1 C_1 E_1) \bar{P}_2$$

$$\text{i.e. } C_1' \bar{P}_2 = (I_1 - \bar{P}_1 C_1 E_1) \bar{P}_2$$

$$\text{i.e. } C_1' = I_1 - \bar{P}_1 C_1 E_1$$

$$\{ (I_1 + E_1 \bar{P}_1)^{-1}\}' = I_1 - \bar{P}_1(I_1 + E_1 \bar{P}_1)^{-1} E_1 \qquad \cdots \ (A4.12)$$

Since E_1 and P_1 are symmetric

$$\{(I_1 + E_1 \bar{P}_1)^{-1}\}' = (I_1 + \bar{P}_1 E_1)^{-1}$$

Using the above relationship in (A4.12),

$$(I_1 + \bar{P}_1 E_1)^{-1} = I_1 - \bar{P}_1 (I_1 + E_1 \bar{P}_1)^{-1} E_1$$

Multiplying both sides by $(I_1 + \bar{P}_1E_1)$

$$I_1 = (I_1 - \bar{P}_1 (I_1 + E_1 \bar{P}_1)^{-1} E_1) (I_1 + \bar{P}_1 E_1)$$

$$= I_1 - \bar{P}_1(I_1+E_1\bar{P}_1)^{-1} E_1 + \bar{P}_1E_1 - \bar{P}_1(I_1+E_1\bar{P}_1)^{-1} E_1\bar{P}_1E_1$$

$$= (I_1+\bar{P}_1E_1) - \bar{P}_1(I_1+E_1\bar{P}_1)^{-1} (I_1+E_1\bar{P}_1) E_1$$

$$= I_1$$

Thus (A4.12) is established.

Using (A4.12) in (A4.11),

$$\bar{F}_1 = \bar{P}_1 C_1 - h\ (\bar{P}_1 C_1 E_2 \bar{P}_2' C_1 + C_1' \bar{P}_2 E_2 \bar{P}_1 C_1) + 0(h^2)$$

$$\bar{F}_2 = h\ C_1'\ \bar{P}_2 + 0(h^2) \qquad \qquad \cdots \quad (A4.13)$$

$$\bar{F}_3 = 0(h^2)$$

Where $C_1 = (I_1 + E_1\ \bar{P}_1)^{-1}$

Similarly, it is shown that G can be expressed as power series in h.

$$G = WC - \bar{P}\ CEWC - h^{N-k}WCEWC$$

$$= [I - (\bar{P} + h^{N-k}W)\ CE]\ WC \qquad \qquad \cdots \quad (A4.14)$$

where $C = [I + E\ \bar{P}]^{-1}$

It is shown in (A4.8) that C is expressable as a power series in h. Thus it follows from (A4.14) that G is also expressable as a power series in h.

From (A4.14), it is clear that the terms associated with h^{N-k} are of first order for a $k = N-1$, second order for $k = N-2$ and so on and so forth. While writing the zeroth and higher order terms for this equation, this aspect should be borne in mind. For zeroth order approximation

$$G^{(0)} = \begin{bmatrix} G_1^{(0)} & G_2^{(0)} \\ \\ G_2^{(0)} & G_3^{(0)} \end{bmatrix} = \left\{ \begin{matrix} I_1 & 0 \\ \\ 0 & I_2 \end{matrix} \right\} - \left\{ \begin{matrix} \bar{P}_1 & 0 \\ \\ 0 & 0 \end{matrix} \right\} \left\{ \begin{matrix} C_1 & 0 \\ \\ -E_2'\bar{P}_1 C_1 & 0 \end{matrix} \right\} \left\{ \begin{matrix} E_1 & E_2 \\ \\ E_2' & E_3 \end{matrix} \right\}$$

$$\begin{bmatrix} W_1 & W_2 \\ W_2' & 0 \end{bmatrix} \begin{bmatrix} C_1 & 0 \\ -E_2'\bar{P}_1 C_1 & 0 \end{bmatrix}$$

where $C_1 = [I_1 + E_1 \bar{P}_1]^{-1}$

or

$$G_1^{(0)} = C_1[W_1C_1 - W_2E_2'\bar{P}_1C_1] - \bar{P}_1 \, C_1 \, E_2 \, W_2' \, C_1$$

$$G_2^{(0)} = C_1' \, W_2$$

$$G_3^{(0)} = 0$$

On similar lines the first and higher order terms of G can be evaluated.

REFERENCES

1. W. Wasow, Asymptotic Expansions for Ordinary Differential Equa-
    tions, Wiley - Interscience, New York, 1965.

2   V. F. Butuzov, A. B. Vasileva, and M. V. Fedoruk, "Asymptotic
    methods in the theory of ordinary differential equations," in
    Progress in Mathematics, R. V. Gamkrelidze, Ed., Plenum, New York,
    pp. 1-82, 1970.

3   W. Eckhaus, Asymptotic Analysis of Singular Perturbations, North-
    Holland, Amsterdam, 1979.

4. A. H. Nayfeh, Introduction to Perturbation Techniques, John Wiley
    & Sons, New York, 1981.

5. W. Eckhaus and E. M. de Jager, (Edrs.), Theory and Applications of
    Singular Perturbations, Lecture Notes in Math., Vol. 942,
    Springer-Verlag, Berlin, 1982.

6. P. V. Kokotovic, "Applications of singular perturbation techniques
    to control problems," SIAM Review, Vol. 26, pp. 501-550, 1984.

7. J. B. Cruz, Jr., (Ed.), Feedback Systems, McGraw-Hill, New York,
    1972.

8. R. E. O'Malley, Jr., Introduction to Singular Perturbations,
    Academic Press, New York, 1974.

9. P. V. Kokotovic, R. E. O'Malley, Jr., and P. Sannuti, "Singular
    perturbations and order reduction in control theory-an overview,"
    Automatica, Vol. 12, pp. 123-132, 1976.

10. M. D. Ardema, (Ed.), Singular Perturbations in Systems and Con-
    trol, CISM Courses and Lectures, No. 280, Springer-Verlag, Wien,
    1983.

11. V. R. Saxena, J. O'Reilly, and P. V. Kokotovic; "Singular Pertur-
    bations and time-scale methods in control theory: survey
    1976-1983," Automatica, Vol. 20, pp. 273-293, 1984.

12. P. Dorato, and A. H. Levis, "Optimal linear regulators: the
    discrete-time case," IEEE Trans. on Aut. Control, Vol. AC-16,
    pp. 613-620, 1971.

13. F. B. Hildebrand, Finite Difference Equations and Simulations,
    Prentice Hall, Englewood Cliffs, 1968.

14. J. A. Cadzow, and H. R. Martens, Discrete-Time and Computer Con-
    trol Systems, Prentice Hall, Englewood Cliffs, 1970.

15. B. C. Kuo, Digital Control Systems, SRL Publ. Comp., Champaign,
    1977.

16. J. A. Cadzow, Discrete-Time System: An Introduction with Inter-
    disciplinary Applications, Prentice Hall, Englewood Cliffs, 1973.

17. A. B. Bishop, Introduction to Discrete Linear Controls: Theory
    and Application, Academic Press, New York, 1975.

18. C. Comstock, and G. C. Hsiao, "Singular perturbations for differ-
     ence equation," Rocky Mountain J. Mathematics, Vol. 6,
     pp. 561-567, 1976.

19. A. Locatelli, and N. Schianoni, "Two-time-scale discrete systems,"
     First Int. Conf. on Inf. Sciences and Systems, Patres, Greece,
     Aug. 1976.

20   F. C. Hoppenstead, and W. L. Miranker, "Multitime methods for sys-
     tems of difference equations," Studies in Appl. Math., Vol. 56,
     pp. 273-289, 1977.

21. H. J. Reinhardt, "Stability of singularly perturbed linear differ-
     ence equations," in Numerical Analysis of Singular Perturbation
     Problems, P. W. Hemker, and J. J. H. Miller, Edrs., Academic
     Press, New York, pp. 485-492, 1979.

22. R. Subramanyam, and A. Krishnan, "Nonlinear discrete systems anal-
     ysis by multiple time perturbation techniques," J. of Sound and
     Vibration, Vol. 63, pp. 325-335, 1979.

23   S. H. Javid, "Multi-time methods in order reduction and control of
     discrete systems," 13th Asilomer Conf. Circuits, Systems, and Com-
     puters, Pacific Grove, CA, 5-7, Nov. 1979.

24. R. G. Phillips, Two-Time-Scale Discrete Systems, M.S. Thesis,
     Report R-839, Coord. Sci. Lab., Univ. of Illinois, Urbana, 1979.

25. R. G. Phillips, "Reduced order modelling and control of two-time-
     scale discrete systems," Int. J. of Control, Vol. 31, pp. 765-780,
     1980.

26. P. K. Rajagopalan, and D. S. Naidu, "A singular perturbation
     method for discrete control systems," Int. J. of Control, Vol. 32,
     pp. 925-936, 1980.

27. R. Atluri, and Y. K. Kao, "Sampled-data control of systems with
     widely varying time constants," Int. J. of Control, Vol. 33,
     pp. 555-564, 1981.

28. G. L. Blankenship, "Singularly perturbed difference equations in
     optimal control problems," IEEE Trans. on Aut. Control,
     Vol. AC-26, pp. 911-917, 1981.

29. D. S. Naidu, and A. K. Rao, "A singular perturbation method for
     initial value problems with inputs in discrete control systems,"
     Int. J. of Control, Vol. 33, pp. 953-965, 1981.

30. A. K. Rao, and D. S. Naidu, "Singularly perturbed boundary value
     problems in discrete systems," Int. J. of Control, Vol. 34,
     pp. 1163-1173, 1981.

31. D. S. Naidu, and A. K. Rao, "A singular perturbation method for
     boundary value problems in discrete systems," IFAC Symposium on
     Theory and Applications of Digital Control, New Delhi, 5-7 Jan.
     1982.

32. D. S. Naidu, and A. K. Rao, "Singular perturbation mehtods for a
     class of initial and boundary value problems in discrete systems,"
     Int. J. Control, Vol. 36, pp. 77-94, 1982.

33. A. K. Rao, and D. S. Naidu, "Singular perturbation method applied to open-loop discrete optimal control problem," Optimal Control: Applications & Methods, Vol. 3, pp. 121-131, 1982.

34. D. S. Naidu, and A. K. Rao, "Singular perturbation analysis of the closed-loop discrete optimal control problem," Optimal Control: Applications & Methods, Vol.5, pp. 19-28, 1984.

35. W. L. Miranker, "The computational theory of stiff differential equations," Lecture Notes, No. 219-7667, V-Paris XI, U.E.R., Mathematique, 91405, Orsay, France, 1975.

36. L. Abrahamson, On Difference Approximations for Singular Perturbations of Ordinary Differential Equations, Ph.D. Dissertation, Uppsala University, Uppsala, Sweden, 1976.

37. J. E. Flaherty, and R. E. O'Malley, Jr. "The numerical solution of boundary value problems for stiff differential equations," Mathematics of Computation, Vol. 31, pp. 66-93, 1977.

38. H. O. Kreiss, "Difference methods for stiff ordinary differential equations," SIAM J. Numerical Analysis, Vol. 15, pp. 21-58, 1978.

39. P. W. Hemker, and J. J. H. Miller, (Edrs), Numerical Analysis of Singular Perturbation Problems, Academic Press, New York, 1979.

40. P. k Rajagopalan, and D. S. Naidu, "Singular perturbation method for discrete models of continuous systems in optimal control," IEE Proc. Part D, Control Theory and Appl., Vol. 128, pp. 142-148, 1981.

41. A. P. Sage, and C. C. White, Optimum Systems Control, Second Edition, Prentice Hall, Englewood Cliffs, 1977.

42. K. Ogata, State Space Analysis of Control Systems, Pentice Hall, Englewood Cliffs, 1967.

43. M. Jamshidi, "Application of three-time-scale near-optimum design to control systems," Automatic Control Theory and Applications, Vol. 4, pp. 7-13, 1976.

44. S. M. Roberts, and J. S. Shipman, Two-Point Boundary Value Problems: Shooting Methods, Elsevier, New York, 1972.

45. E. A. Guillemin, The Mathematics of Circuit Analysis, Oxford & IBH Publ. Col, Calcutta, 1967.

46. V. Dragan, "Hybrid control for systems with small time constants," Rev. Roum. Sci. Techn. Elect. et. Energ., Vol. 25, pp. 289-303, 1980.

47. V. Dragan, "Hybrid control of singularly perturbed stationary systems," Rev. Roum. Sci. Techn. Elect. Energ., Vol. 25, pp. 617-626, 1980.

48. A. Bradshaw, and B. Porter, "Singular perturbation methods in the design of tracking systems incorporating fast sampling error-activated controllers," Int. J. Syst. Sci., Vol. 12, pp. 1181-1191, 1981.

49. A. Bradshaw, and B. Porter, "Singular perturbation methods in the design of tracking systems incorporating inner-loop compensators and fast sampling error actuated controllers," Int. J. Syst. Sci., Vol. 12, pp. 1207-1220, 1981.

50. F. Delebecque, and J. P. Quadrat, "Optimal control of Markov chains admitting strong and weak interaction," Automatica, Vol. 17, pp. 281-296, 1981.

51. V. Dragan, "A new case in the hybrid control of non-stationary systems with fast transients," Rev. Roum. Sci. Tech. Elect. et. Energ., Vol. 26, pp. 447-453, 1981.

52. V. Dragan, "Suboptimality results in the hybrid control of singu- larly perturbed stationary systems," Rev. Room Sci. Techn. Elect. et. Energ., Vol. 26, pp. 587-593, 1981.

53. M. S. Mahmoud, and M. G. Singh, Large-Scale Systems Modelling. Pergamon Press, Oxford, 1981.

54. B. Porter, "Fast-sampling tracking systems incorporating Lur'e plants with multiple nolinearities," Int. J. Control, Vol. 34, pp. 345-358, 1981.

55. K. V. Fernando, and H. Nicholson, "Singular perturbation model reduction in the frequency domain," IEEE Trans. Aut. Cont., Vol. AC-27, pp. 969-970, 1982.

56. P. A. Ioannou, Robustness of Model Reference Adaptive Schemes with Respect to Model Errors, Ph.D. Thesis, Rep. R-955, Coordinated Science Lab., Univ. of Illinois, Urbana, USA, 1982.

57. M. S. Mahmoud, "Design of obsever-based controllers for a class of discrete systems," Automatica, Vol. 18, pp. 323-328, 1981.

58. M. S. Mohmoud, "Order reduction and control of discrete systems," IEE Proc. Part D, Control Theory & Appl., Vol. 129, pp. 129-135, 1982.

59. M. S. Mahmoud, "Structural properties of discrete systems with slow and fast modes," Large Scale Systems, Theory Appl., Vol. 3, pp. 227-236, 1982.

60. A. K. Rao, Singular Perturbation Analysis of Difference Equations with Applications to Control Problems, Ph.D. Thesis Dept. of Elect. Engy., Indian Inst. of Technology, Kharagput, 1982.

61. F. Delebeque, "A reduction process for perturbed Markov chains," SIAM J. Appl. Math, Vol. 43, pp. 324-350, 1983.

62. K. V. Fernando and H. Nicholson, "Singular perturbation approxima- tions for discrete time balanced systems," IEEE Trans. Aut. Con- trol, Vol. AC-28, pp. 240-242, 1983.

63. P. A. Ioannou and C. R. Johnson, Jr., "Reduced order performance of parrallel and series parallel identifiers with weakly observ- able parasitics," Automotica, Vol. 19, pp 75-88, 1983.

64. P. A. Ioannou and P. V. Kokotovic, Adaptive Systems with Reduced Models, Lect. Notes in Control and Inf. Sciences, Vol. 47, Springer Verlag, Berlin, 1983.

65. H. Kando and T. Iwazumi, "Suboptimal control of discrete regulator problems via time-scale decomposition, Int. J. Control, Vol. 37, pp 1323-1347, 1983.

66. H. Kando and T. Iwarzumi, "Initial value problems of singularly perturbed discrete systems via time-scale decomposition," Int. J. Syst. Sci., Vol. 14, pp. 555-570, 1983.

67. M. Kimura, "On the matrix Riccati equation for a singularly perturbed linear discrete control system," Int. J. Control, Vol. 38, pp. 959-975, 1983.

68. B. Litkouhi, and H. K. Khalil, Multirate and composite control of two-time-scale discrete time systems," in Singular Perturbation Methods and Multimodel Control, H. K. Khalil (Ed.), Final Report pp. 136-196, Michigan State Univ., East Lansing, December 1983.

69. R. G. Phillips, "The equivalence of time-scale decomposition techniques used in the analysis and design of linear systems," Int. J. Control, Vol. 37, pp. 1239-1257, 1983.

70. G. P. Syrcos, and P. Sannuti, "Singular perturbation modeling of continuous and discrete physical models," Int. J. Control, Vol. 37, pp. 1007-1022, 1983.

71. M. T. Tran, and M. E. Sawan, "Reduced order discrete models," Int. J. Syst. Sci., Vol. 16, pp. 745-752, 1983.

72. M. T. Tran, and M. E. Sawan, "Nash strategies for discrete-time systems with slow and fast modes," Int. J. Syst. Sci., Vol. 16, pp. 1355-1371, 1983.

73. B. Litkouhi, and H. Khalil, "Infinite-time reulator for singularly perturbed difference equations," Int. J. Control, Vol. 39, pp. 587-598, 1984.

74. A K. Rao, and D. S. Naidu, "Singular perturbation method for Kalman filter in discrete systems," IEE Proc. Control Theory & Appl., Part D, Vol. 131, pp. 39-46, 1984.

75. M. T. Tran and M. E. Sawan, "Low order observer for discrete systems with slow and fast modes," Int. J. Syst. Sci, Vol. 15, pp. 1283-1288, 1984.

SUBJECT INDEX